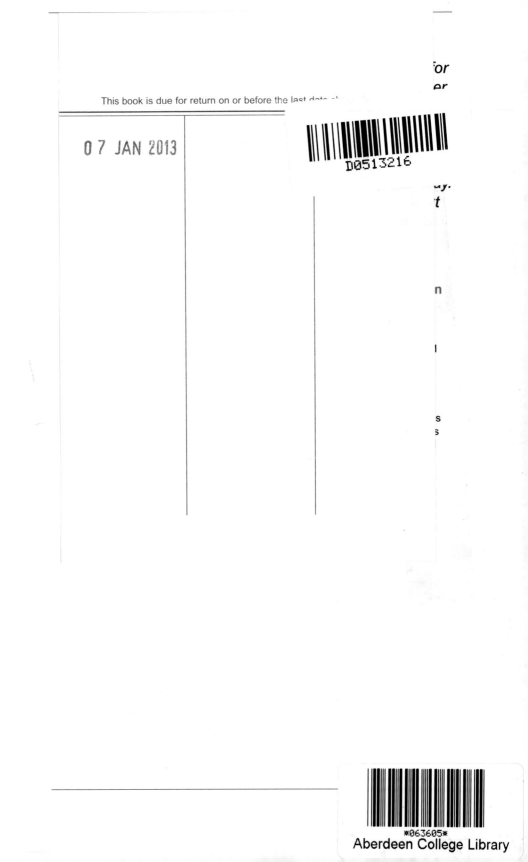

This book is due for return on or before the last date shown below

07 JAN 2013

D0513216

safety

sharing**the**experience

improving the way lessons are learned
through people, process and technology

BP Process Safety Series

Hazards of Electricity and Static Electricity

**A collection of booklets
describing hazards and
how to manage them**

bp

This booklet is intended as a safety supplement to operator training courses, operating manuals, and operating procedures. It is provided to help the reader better understand the 'why' of safe operating practices and procedures in our plants. Important engineering design features are included. However, technical advances and other changes made after its publication, while generally not affecting principles, could affect some suggestions made herein. The reader is encouraged to examine such advances and changes when selecting and implementing practices and procedures at his/her facility.

While the information in this booklet is intended to increase the store-house of knowledge in safe operations, it is important for the reader to recognize that this material is generic in nature, that it is not unit specific, and, accordingly, that its contents may not be subject to literal application. Instead, as noted above, it is supplemental information for use in already established training programmes; and it should not be treated as a substitute for otherwise applicable operator training courses, operating manuals or operating procedures. The advice in this booklet is a matter of opinion only and should not be construed as a representation or statement of any kind as to the effect of following such advice and no responsibility for the use of it can be assumed by BP.

This disclaimer shall have effect only to the extent permitted by any applicable law.

Queries and suggestions regarding the technical content of this booklet should be addressed to Frédéric Gil, BP, Chertsey Road, Sunbury on Thames, TW16 7LN, UK. E-mail: gilf@bp.com

Published by
Institution of Chemical Engineers (IChemE)
Davis Building
165–189 Railway Terrace
Rugby, CV21 3HQ, UK

IChemE is a Registered Charity
Offices in Rugby (UK), London (UK) and Melbourne (Australia)

© 2005 BP International Limited

ISBN 0 85295 474 3

First edition 1961; Second edition 1962; Third edition 1964; Fourth edition 1983; Fifth edition 2005

Typeset by Techset Composition Limited, Salisbury, UK
Printed by Henry Ling, Dorchester, UK

Foreword

Electricity, like fire, can either be the best friend or worst enemy of anyone who uses it. This booklet is intended to raise the awareness of those operators, engineers and technicians working on process plant of how static electricity could be unexpectedly generated or electricity misused, in order to adopt safe designs and practices to avoid the occurrence of such incidents.

I strongly recommend you take the time to read this book carefully. The usefulness of this booklet is not limited to operating people; there are many useful applications for the maintenance, design and construction of facilities.

Please feel free to share your experience with others since this is one of the most effective means of communicating lessons learned and avoiding safety incidents in the future.

Greg Coleman, Group Vice President, HSSE

W.H. Colen

Acknowledgements

The co-operation of the following in providing data and illustrations for this edition is gratefully acknowledged:

- BP Refining Process Safety Network
- Jim Bickerton, IChemE Loss Prevention Panel member

Contents

1   Introduction 1

2   Sparks, arcs and ignition energy 2

3   Area classification 4

4   Hazards of static electricity 8
4.1 General considerations 8
4.2 Specific rules and applications 15
4.3 The legend of mobile phones 27

5   Hazards of lightning 28

6   Hazards of stray currents 33

7   Bonding and grounding 38
7.1 For dissipation of static charges 38
7.2 For protection of personnel and equipment 40

8   Hazards of electrical shock 44
8.1 Rules for prevention of electrical shock 45
8.2 Rules for minimizing injury from electrical shock 46

9   Explosion and fire hazards of electrical equipment 48
9.1 Batteries 55
9.2 Electrical fires 56
9.3 Fire protection of electrical/instrumentation/computer rooms . . . 57

10 Dangers of improper operation of electrical equipment . . . 58

11 Safeguards for electrical equipment **61**
11.1 General 61
11.2 Contact with overhead cables 63
11.3 Excavation work 68
11.4 Training and safe working practices 69
11.5 Identification of disconnecting means and circuits 72
11.6 Re-energizing equipment 73
11.7 Working with energized equipment 74

12 Power outages **76**
12.1 General 76
12.2 The regulator point of view 78
12.3 Triggering events 78
12.4 What can be done to keep plants safe in case of a
       power outage? 80

13 Some points to remember **81**

     Appendix 1. Diesel driven emergency equipment **86**

     **Appendix 2. Short bibliography for regulations
     and norms** **93**

     **Appendix 3. Ignition temperatures for common dusts
     and gases/vapours** **95**

     Test yourself! **96**

1

Introduction

In petrochemical plants and refineries, electric sparks and electrical shock are the two principal hazards of electricity. Sparks and arcs may ignite mixtures of air and flammable gases or vapours, resulting in explosions and fires. Electrical shock may cause fatalities or serious injuries. This booklet describes these hazards and suggests preventive measures but it does not replace regulations and procedures.

2

Sparks, arcs and ignition energy

Electric sparks and arcs occur in the normal operation of certain electrical equipment such as switches, brushes and similar devices. They also occur during the breakdown of insulation on electrical equipment.

When electricity jumps a gap in air, it is called a *spark* (Figure 1). We are all familiar with the static spark that may jump from the end of a finger to a metal switchplate after one has walked across a carpet (Figure 2).

Because of electrical inertia (inductance), an arc occurs when two contacts are separated (Figure 3). This inertia, or inductance, simply means that electricity which is flowing tries to keep on flowing, just as a turning flywheel tries to keep on turning. Circuits which contain coils have a large amount of electrical inertia. The amount of current flowing when the contacts are first separated is also of great importance because current helps determine the intensity of the arc.

The minimum amount of energy which a spark or arc must have to ignite a flammable mixture is extremely small.

Most electrical equipment can produce sparks and arcs which have more than enough energy to cause ignition. The requirements for the safe installation of such equipment in the oil industry are discussed in Chapters 3 and 4.

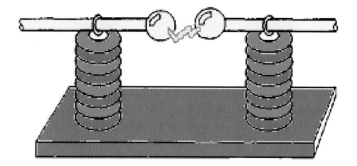

Figure 1 *Electricity jumping across an air gap is called a spark.*

Some electrical devices may create sparks or arcs that are too weak to cause ignition. The switches on flashlights or switches in thermocouple circuits are examples. However, the hot filament of a flashlight bulb can cause ignition if the bulb is broken. Whether a switch in a low-energy circuit will create an arc hot enough to cause ignition will depend upon the circuit characteristics. The switch on a flashlight will not create an arc hot enough to cause ignition because there is little inductance in the circuit. However, the same switch would produce a spark hot enough to cause ignition if the circuit contained an automobile ignition coil.

Therefore, a switch cannot be said to be safe unless the complete characteristics of the circuit in which it is used are known and carefully studied.

Because circuit characteristics vary with each wiring installation and circuits can change by damage or breakdown, there is no easy way to be sure that nonexplosion-proof switches or devices are safe or will remain safe. Remember that it is not the switch in itself but both the switch and its circuits that determine whether or not any given application of a switching device is safe in a hazardous location. Safe application of nonexplosion-proof switches to low-energy circuits located in hazardous areas is a matter that requires careful study. For example, it has been determined that certain transistorized radios are suitable for operation in a hazardous area. Each specific application should be approved by a qualified electrical engineer.

Circuits feeding explosion-proof or vapour-tight lighting fixtures must be turned off before faulty lamps are replaced. Replacing a lamp in a 'hot' circuit will spark and could ignite explosive vapours.

Surface temperatures of electrical devices such as heating cable, hot plates, lamps, electronic tubes, etc., must not exceed 80 percent of the ignition temperature of the gas or vapour likely to be present in the area. These surface temperatures must be determined before equipment is installed.

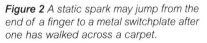

Figure 2 A static spark may jump from the end of a finger to a metal switchplate after one has walked across a carpet.

Figure 3 An arc occurs when current-carrying contacts are separated.

3

Area classification

Many areas in chemical/oil plants are considered hazardous because of the possible presence of flammable gases or vapours. In a few areas, these gases or vapours are present all or most of the time. In most areas, they are present only infrequently and for a short time.

Hazardous areas are of two types, as follows:

- Areas which are considered always hazardous because flammable gases or vapours will be present all or most of the time under normal conditions.
- Areas which are considered hazardous only infrequently as a result of ruptures, leaks or other unusual circumstances.

Likewise, electrical equipment is of two types, as follows:

- Equipment which creates sparks or arcs as a result of normal use. An ordinary light switch is an excellent example of a device that arcs during normal use.
- Equipment which creates sparks or arcs only at the instant of failure. The common induction motor used to drive most refinery pumps is an example of such equipment which normally operates for a number of years without failure.

Obviously if an area is always hazardous, all the electrical equipment in that area must be enclosed in explosion-proof housings. Otherwise, a spark or an arc is sure to start a fire or cause an explosion. (Explosion-proof housings are explained in more detail in Chapter 9.)

But what about areas that are only hazardous infrequently and for a short time? Must all electrical equipment in these areas be explosion-proof?

The answer is that some, but not all, of the electrical equipment in such areas must be explosion-proof. First, consider equipment that sparks or arcs during normal operation. Sooner or later, the infrequent presence of a flammable gas or vapour and one of those normal sparks or arcs will occur at the same time, and a fire or explosion will probably result. To prevent this, electrical equipment which sparks or arcs during normal operation must always be explosion-proof, even if they are used in an area which is only infrequently hazardous.

Now consider electrical equipment which sparks or arcs only at the instant of failure. What chance is there that such a spark or arc will occur during one of the infrequent and short periods when a flammable gas or vapour is present? There is almost no chance that these events will occur at the same time.

Therefore, electrical equipment which sparks or arcs only at the instant of failure need not be explosion-proof when used in areas which are only infrequently hazardous.

Now let us look at the way in which the general ideas discussed above are covered by the specific rules given in the US National Electrical Code (NEC). The NEC classifies hazardous areas of various types and states what sort of equipment is safe for use in each. Refineries are most concerned with those areas which the NEC calls Class I, Group D.

- *Class I* locations are those in which flammable gases or vapours are or may be present in the air in amounts large enough to produce ignitable or explosive mixtures.
- *Group D* refers specifically to atmospheres containing gasoline, hexane, naphtha, benzene, butane, propane, alcohol, lacquer solvent vapours or natural gas.
- Class I locations are further divided into Division 1 and Division 2 locations.

Class I, Division 1 locations are considered always hazardous because flammable gases or vapours exist continuously or intermittently under normal operation, or frequently because of repairs or maintenance. Class I, Division 2 locations are rarely hazardous because flammable liquids or gases are handled in closed containers or piping systems from which they cannot escape except in accidental cases of rupture or breakdown of containers or piping.

Most refinery areas are classified Class I, Group D, Division 2. A few are considered Class I, Group D, Division 1; and some areas, of course, are not hazardous at all. The NEC requires that all electrical equipment in a Class I, Group D, Division 1 location be explosion-proof. In a Class I, Group D, Division 2 area, only equipment which sparks or arcs during normal operation must be explosion-proof.

The table below gives a quick comparison between US and European classifications:

| USA | EU | | Area where an explosive mixture is: |
	Gas	Dust	
Division 1	Zone 0	Zone 20	likely to be present at all times under normal operating conditions.
	Zone 1	Zone 21	likely to occur in normal operation.
Division 2	Zone 2	Zone 22	not likely to be occur in normal operation, and if it does it is only for short periods.

Figures 4, 5 and 6a, based on American Petroleum Institute Bulletin No. RP-500, show typical Division 1 and Division 2 areas.

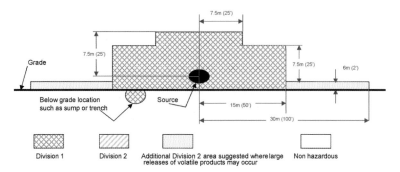

Figure 4 *Typical area classifications in a freely ventilated process area (source of hazard near grade).*

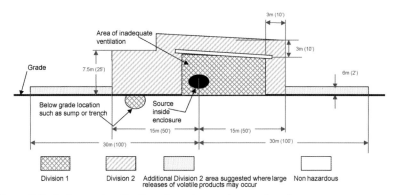

Figure 5 *Typical area classifications in and around an enclosed process area.*

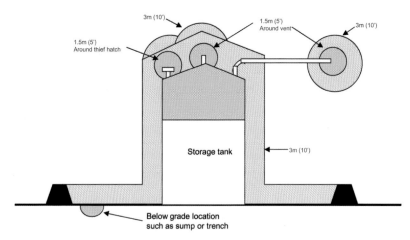

Figure 6a *Typical area classifications in a tank farm.*

Figure 6b based on European EN 60079-10 shows typical Zones 0, 1 and 2 areas for a service station

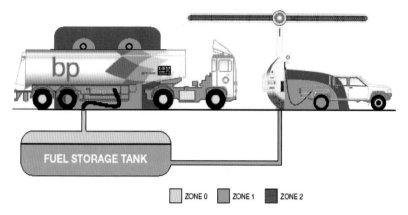

ZONE 0   ZONE 1   ZONE 2

Figure 6b Typical area classifications for a service station (zones shown only on light vehicle side of delivery meter).

4

Hazards of static electricity

4.1 General considerations

The principal hazard of static electricity is a spark discharge which can ignite a flammable mixture. Refined flammable liquids—such as gasoline, kerosene, jet fuels, fuel oils and similar products—become charged with static electricity from pumping, flow through pipes, filtering, splash filling or by water settling through them. Different liquids, of course, generate different amounts of static electricity. Refined hydrocarbon liquid fuels vary widely in their ability to generate and to conduct static electricity. Broadly speaking, the products that are better conductors are also better generators of static; but because they are better conductors, the static electricity generated is discharged much more readily. In most cases the electrostatic charge that is generated in a liquid is released instantaneously to a ground because the liquid is a good conductor. But there are some cases where a charge is accumulated in a liquid because that liquid has a low electrical conductivity. Such a liquid is called a static accumulator.

A static accumulator is defined as a substance which can accumulate an electrostatic charge and retain that charge for significant period of time. Non-conductive substances retain charges because the charges cannot flow through or across the substance to a ground. Ungrounded conductive objects are also static accumulators. Static accumulator liquids are generally found in the more highly refined products.

It is the petroleum products with the *least* conductivity that pose the *greatest* danger. This is due to their inherent inability to dissipate an electrical charge. The International Safety Guide for Oil Tankers and Terminals (ISGOTT) states that, in general, black oils do not accumulate a static charge and clean oils (distillates) do:

Non-accumulator oils	Accumulator oils
Crude oils Black diesel oils Asphalts	Natural gasolines Kerosenes White spirits Motor and aviation gasolines Jet fuels Naphthas Heating oils Clean diesel oils Lubricating oils Residual fuel oils

Conductivity is measured in picosiemens per meter (pS/m) while resistivity is measured in ohm-cm.

Conductivities of some flammable chemicals

Liquid	Conductivity, σ (pS/m)	Liquid	Conductivity, σ (pS/m)
Acetone	5×10^8	Toluene	1
Methyl ethyl ketone	5×10^6	Xylene	0.1
Ethyl benzene	30	Heptane	3×10^{-2}
Styrene monomer	10	Benzene	5×10^{-3}
Cyclohexane	2	Hexane	1×10^{-5}

Note: $5 \times 10^8 = 500,000,000$

$1 \times 10^{-5} = 0.00001$

Electrostatic accumulation is significant *unless**:

- conductivity exceeds 50 pS/m (resistivity less than 2×10^{12} ohm cm);
- product is handled in earthed/grounded conductive containers.

*Not applicable for mists

ACCIDENT A technician was performing a routine 'soap test' on a gasoline product line that had been damaged and repaired. The product line was pressurized with nitrogen, but also contained residual gasoline.

At the conclusion of the test, the technician removed the isolation device from the head of the submersible turbine pump (STP), which allowed the uncontrolled release of gasoline vapours and/or liquid into the air.

The vapours ignited and engulfed the STP manhole and the technician in flames. At the time of the accident, the technician was lying on the concrete above the STP manhole, reaching in to disconnect the testing apparatus.

The technician experienced third degree burns over the upper portion of his body and suffered respiratory damage. He died six days later.

Submersible turbine pump in sump

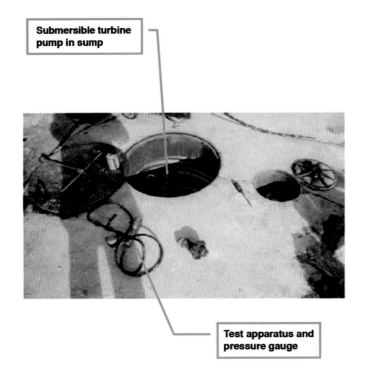

Test apparatus and pressure gauge

The fire resulted from the uncontrolled and rapid release of the line's contents and subsequent ignition of the vapours. Both a consulting specialty engineering firm and a consulting fire forensics professional determined that static electric discharge, caused by the rapid release of product and vapours, was the most likely source of ignition.

Relaxation

The discharge process is called *relaxation*, and relaxation time is often expressed as the time required for a given charge to decrease to half its original value. If this time is very short, large static potentials in the bulk fuel are not created because the relaxation process takes over and controls the charge that can build up.

The poorer conductors (generally the cleaner products) are also poorer generators; but because their relaxation times are so much longer, large charges can be generated. Very poor conductors, though having excessive relaxation times, are also such poor generators that hazardous static potentials might not be created.

Two static hazards may result when refined hydrocarbon liquids are pumped into a tank. The first and most dangerous is the sparking that can occur on the liquid surface. The second is the accumulation of a static charge on the receiving tank if the tank is insulated from ground. Both hazards are shown in Figure 7.

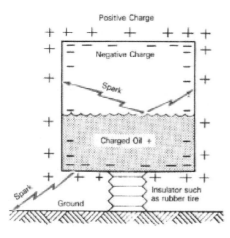

Figure 7 A tank or container insulated from ground (such as a tank truck) may spark across the oil surface, from the shell to the ground or to a conductive object connected to ground (such as a man).

Accumulation of a static charge on the outside of a tank (and therefore a possible spark discharge from the tank shell to a grounded object) will be eliminated if the tank is grounded or is in contact with the earth (Figure 8).

However, grounding a receiving tank does not eliminate the chance of surface sparking inside the tank. The only certain way to prevent this is to eliminate the free surface by using a floating-roof tank.

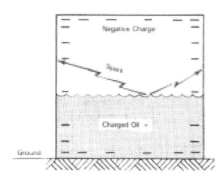

Figure 8 Charge on the outside of the shell will drain to earth when the tank or container is grounded, leaving the negative charge on the inside of the shell. This grounding can be done with a wire (a tank-truck ground, for example), or the tank or container may rest on the ground. However, grounding the shell does not eliminate the chance of sparking inside the tank from the charged oil to internal structural members or to the tank shell.

This also has the advantage of eliminating the vapour space above the liquid. Static charge in an ordinary tank (and hence the chance of surface sparking) will decrease to a safe level during the relaxation time. Relaxation times are normally a matter of seconds, but they may range up to two hours.

Static sparks on a free hydrocarbon liquid surface inside of a tank may occur as the result of the flow of charged oil into the tank, charge generated by splash and spray from the inlet stream, or both (Figure 9). The inlet stream becomes charged as the liquid flows through filters, pumps and piping; and the charge increases as the flow velocity increases. High flow velocities may also increase static generation by increasing splashing and spraying inside the tank. Therefore, the possibility of sparks on the free liquid surface in the tank can be reduced by decreasing the flow velocity to and into the tank.

Several variable factors must be favourable for the production of both a static spark and a flammable vapour-air mixture. The probabilities are very much against these factors combining to cause fires or explosions, but occasionally they will do so. Two of the factors, the velocity of flow and the vapour-air mixture, may be controlled to reduce the ignition hazard. Flow restricted to velocities of about 3 feet per second (1 m/s) or less will reduce static generation.

Vapour spaces can be made inert with steam, nitrogen, flue gas or carbon dioxide (refer to the BP Process Safety Booklets *Hazards of Steam* and *Hazards of Nitrogen and Catalyst Handling* for more details).

Non-volatile combustible liquids can produce mists, which are easily ignitable even though the temperature is well below the flash point. Fine mist droplet build-up and splash filling can create an electrostatic discharge sufficient to act as a source of ignition. The nitrogen purge and inlet/return stilling tubes in a gas compressor's seal oil tank are critical safety systems, which must be reinstated after maintenance work.

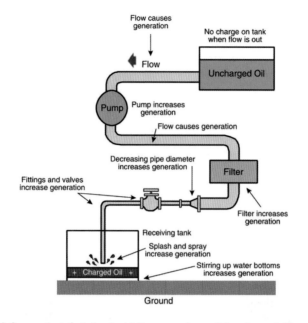

Figure 9 *Generation of static electricity occurs from oil flow through filters, pumps, pipe and fittings, and by splashing and spraying. Since flow in a pipe generates static electricity at a relatively low rate, increasing the length of straight pipe downstream from a filter or other high static-charge generator can provide additional relaxation time to reduce the high static charge. For more information on static generation during transfers, refer to the BP Process Safety Booklet* **Tank farm and (un)loading safe operations.**

In his book *Myths of the chemical industry*, Trevor Kletz describes how the misuse of small containers and static accumulator liquids is potentially very hazardous, under the title 'Which of matches or plastic buckets are more dangerous on a plant handling flammable liquids?'

ACCIDENT A fire occurred at a process plant when the discharge from a drain valve was collected in a metal bucket suspended from the drain valve body. The bucket was not effectively earthed/grounded since there was a thermoplastic sleeve around its handle preventing metal-to-metal contact. The assembly was also some 3 metres (10 ft) above ground level.

Oil had sprayed from the drain valve (open about 20%) for less than a minute when ignition took place. The flames reached about 3 metres (10 ft) high. The fire was confined to the drain bucket, and was quickly extinguished and the plant shut down. There was no re-ignition and the fire caused only superficial damage.

The most likely cause for ignition was a static discharge, arcing from the metal bucket to the drain nozzle; the resulting spark being of sufficient energy to ignite the gaseous mist produced by the high pressure spray from the drain valve.

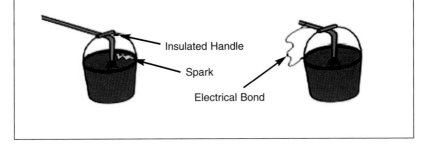

Insulated Handle

Spark

Electrical Bond

A typical illustration of fires started by a spark of static electricity during flammable liquid transfers is found in fires at gasoline service stations involving the filling of fuel containers.

Gasoline has a low electrical conductivity. In other words, it does not conduct electricity very well. As a result, a charge of static electricity builds up on gasoline as it flows through a pipe or hose and this charge takes several seconds to several minutes to dissipate after the gasoline has reached a container. If this charge is allowed to build up enough it may discharge as a spark from a container to a grounded metal object such as a pipe flange, drain nozzle, bucket handle, etc, and it may ignite the gasoline. Ignition requires the spark to occur where the gasoline vapour is in the flammable range.

The condition most likely to lead to a spark discharge is splash filling an insulated or ungrounded container. This is the situation that exists when a metal container is placed on a plastic bedliner of a pickup truck or on the carpet of a car trunk (see 1998 alert from US National Institute for Occupational Safety and Health (NIOSH)). The insulating effect of the bedliner or carpet prevents the static charge generated by gasoline flowing into the container or other sources from grounding. Metal containers, when grounded, provide the greatest protection against fires caused by static electricity.

Greatest hazard	Ungrounded metal container
Less hazard	Non-conducting container (e.g., plastic container)
Least hazard	Grounded metal container

Regulating the size of the container is a good way of ensuring that the static charge will not build up to reach discharge level. Bonding/grounding plastic cans is ineffective so the amount of charge they are allowed to accumulate is controlled by limiting the physical dimensions instead. Bonding/grounding metal cans up to this size is not required but larger cans will require it.

Plastic cans should be limited to 5 litres for forecourt operations (service station), but refineries usually use a lower limit of 1 litre (1 litre = 1.06 quart) . Metal cans up to 10 litres are generally considered adequate if the following precautions are followed:

- the can and person filling it should be standing on the floor with the appropriate ground resistance (<108 ohms);
- no filling in the back of trucks/cars;
- funnels should not be used;
- the dispenser nozzle should be kept well inserted and in contact with the can mouth whilst filling .

Above these sizes, more drastic precautions should be enforced (such as low filling rate, grounding).

4.2 Specific rules and applications

Loading tank cars and tank trucks

Now that we have finished a general discussion of the static electricity problem, let us look at some specific oil industry situations, such as the loading of tank cars and tank trucks. First of all, let us discuss some of the products to be loaded, and how vapour spaces above the liquid become filled with flammable mixtures during loading.

Gasoline vaporizes readily and usually produces vapour-space mixtures with air which are too rich to be flammable. If gasoline is loaded into a truck which contains enough air, the vapour-space mixture will pass through the flammable range. Wind blowing into a loading hatch can also dilute the gasoline vapours to produce a flammable mixture.

Jet fuels, benzene, toluene and special naphthas are particularly dangerous.These products do not vaporize as readily as gasoline, and a flammable vapour-space mixture will always be present under normal loading conditions.

Products having a high flash point such as kerosene, No. 1 fuel oil. No. 2 fuel oil, diesel fuel and lube oil may also be hazardous to load. These products in themselves do not vaporize enough to produce flammable mixtures in the vapour space, so everyone thinks that all is well. The previous load of product having a low flash point (such as gasoline) is forgotten. Then the high flash point product being loaded absorbs some of the rich vapours from the previous load, and the vapour space can soon contain a flammable mixture. This, coupled with the fact that most highly refined products having a high flash point are excellent generators of static electricity, combines all the elements necessary for an explosion and fire.

Consequently, you must think about what the last person loaded as well as about what you are loading. *The loading of a product having a high flash point after a load of product having a low flash point (sometimes called a high proportion of switch loading) is the principal factor in loading-rack truck fires.*

Loading procedures used to reduce static generation from splash and spray will also decrease mist formation. Like a vapour, a fine mist can be an explosion hazard when loading products such as kerosene.

Here are some *do's* and *dont's* for loading tank trucks with products of the types discussed above:

Do

- *Do* be sure the loading-rack grounding connections are in good condition.
- *Do* ground tank trucks immediately after the truck is spotted for loading and before the loading spout is inserted, as shown in Figure 10. The connection shown eliminates the chance of any spark discharge from the truck to the loading pipe or other grounded object.

  Remember, however, that grounding the tank truck does not eliminate the chance of a spark discharge from the oil surface inside the tank.

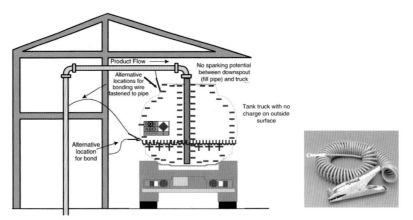

Figure 10 Ground tank trucks immediately after the truck is spotted and before the loading spout is inserted.

- *Do* inspect the interior of the tank compartments. Make sure they are completely drained if the product to be loaded is different from the previous load. Remove all loose foreign objects. Any object that can float or that can be dislodged by product flow during loading may promote sparking. *Note: if entry is required, make sure that good confined space entry rules are adhered to.*

- *Do* inert tank compartments with CO_2 or another inert gas; or purge with airjet, steam-jet or other eduction equipment, such as that shown in Figure 11 for protection against switch loading accidents. The eduction equipment removes rich, low flash point vapours before loading high flash point products. Switch loading is currently responsible for nine out of ten loading-rack truck fires because the high flash point product absorbs some of the rich low flash point vapours, creating an explosive mixture in the compartment, and also the high flash point product is an excellent static generator.

Figure 11 Switch loading accidents can be prevented by purging flammable gases with an eductor.

- *Do* be sure that the tank truck is suitable for the product you wish to load. If it is necessary to flush the tank, the wash stream should enter slowly and should not be jetted or sprayed.

- *Do* extend top-entering loading spouts to the bottom of the tank truck being loaded, as shown in Figure 12.

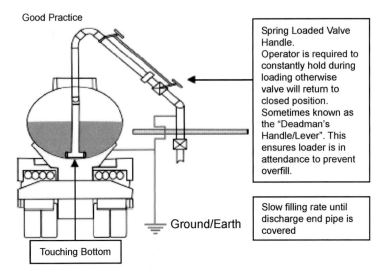

Good Practice

Spring Loaded Valve Handle. Operator is required to constantly hold during loading otherwise valve will return to closed position. Sometimes known as the "Deadman's Handle/Lever". This ensures loader is in attendance to prevent overfill.

Ground/Earth

Slow filling rate until discharge end pipe is covered

Touching Bottom

Figure 12 A top-entering loading spout should extend to the bottom of the tank truck being loaded.

- *Do* use a proper type of deflector on the loading spout, as shown in Figure 13. A good deflector helps prevent the spout from being thrown out of the tank, and when resting on the tank bottom minimizes the spraying and splashing which create mists and help generate a static charge. A tee on the end of a loading spout, as shown in Figure 14, is not satisfactory because it causes excessive splash and spray which increase the generation of static electricity.

Figure 13 A simple deflector resting on the tank bottom reduces splashing and spraying and helps counter reaction force.

Courtesy of OPW Division of Dover Corp.

18

- For products other than gasoline, *do* use a rate of loading low enough to minimize splash and spray until the end of the loading spout is submerged.

Figure 14 A tee on the end of a loading spout should not be used because it produces splash and spray which add to the generation of static electricity.

Don't

- *Don't* start to load a high flash point product such as kerosene until you have drained the truck and made sure that the previous load was a product having a high flash point. If a low flash point product such as gasoline was in the previous load, do whatever is required to make the truck safe to receive the product having a high flash point. An air-jet eductor is one device that has been used to remove low flash point product vapours. *Switch loading (the loading of a product having a high flash point after a load of product having a low flash point) is the principal factor in a high proportion of loading-rack truck fires (Figures 15 and 16).*

Figure 15 This tank-truck explosion and resultant fire were attributed to switch loading and static electricity. Diesel fuel was being loaded into a truck which had hauled gasoline in the previous load. The insert shows a piece of the truck tank which was blown away by the explosion.

Figure 16 Switch loading and static electricity combine to destroy again! This explosion and fire occurred when No. 2 fuel oil was being loaded into a truck which last carried gasoline.

- *Don't* keep a top-entering loading spout high, as shown in Figure 17. If you do, the splash and spray will help build up a static charge, and a spark discharge between the rising liquid surface and the end of the spout may occur.

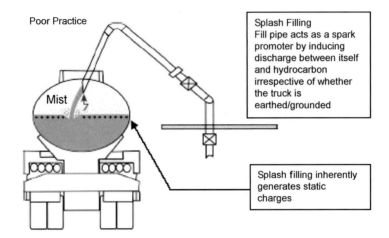

Figure 17 *Splash and spray from a high spout will help generate a static charge, and a spark discharge from the liquid surface to the spout may be a source of ignition.*

- *Don't* withdraw a loading spout from the tank bottom immediately after loading. Wait at least a minute for surface charge to relax. Otherwise, there may be a spark discharge from the liquid surface to the loading spout, as shown in Figure 18.

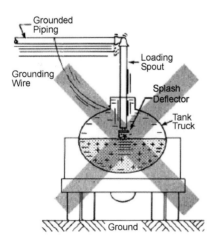

Figure 18 *Wait at least a minute after loading before withdrawing the spout so that surface charge can relax. Otherwise, a spark discharge from the liquid surface to the spout may be a source of ignition.*

- *Don't* sample by dipping through the top opening during loading or immediately after loading. Wait at least a minute for surface charge to relax. Spark discharges between the liquid surface and dipper or thief, or between the charged dipper or thief and the tank rim, have occurred because someone was in a hurry to obtain a sample.

The *do's* and *don'ts* for loading tank cars are similar to those given above for tank trucks. However, no special grounding is required for tank cars because the railroad tracks are used to provide the ground connection. It is important, though, to be sure that the rails are bonded to the loading pipes to prevent stray-current arcs between the edge of the dome and the fill pipe.

Loading tankers and barges

Loading tankers and barges is hazardous for the same reasons that loading tank cars and trucks is hazardous, and similar safety rules should be followed. There are a few special precautions to be taken, however.

During the 1980s and early 1990s, product tankers and tank barge explosions in which static discharge was a probable cause refocused attention on the mechanisms of electrostatic discharge and the applicable safety standards. In most cases, routine cargo tank operations such as loading, stripping or cleaning were underway when these accidents occurred.

When loading, surface charge and surface sparking can be reduced by decreasing flow velocity, as previously discussed. Therefore, ships and barges are often loaded slowly by gravity flow, at least until the inlet on the compartment bottom is covered by a foot or two of liquid. This also eliminates any static charge which might have been produced by splashing and spraying and reduces the risk of forming flammable mists or sprays.

The hazards of sampling by dipping through a top compartment opening are the same as those for tank cars and trucks. Wait after filling and before taking samples so that surface charge can relax. The same safety rules apply to gauging through open hatches.

However, gauging inside a gauging tube can be done at any time because the pipe prevents a spark discharge between the liquid surface and the end of the gauge rod or bob.

For positive protection when loading tankers and barges, it is necessary to eliminate flammable mixtures.

Blanketing the product surface with carbon dioxide or purging the entire compartment with scrubbed flue gas are two of the methods used.

ACCIDENT An explosion occurred in No.2 cargo tank of a barge while it was being loaded with benzene. Although the explosion caused considerable damage to the vessel, there were no injuries. The investigation carried out by the Netherlands Organization for Applied Scientific Research (TNO) concluded that the direct cause of the explosion was a flammable aerosol/air atmosphere that was ignited by a discharge of static electricity during initial filling.

Prior to loading a new chemical, the cargo tanks are completely emptied and gas freed. Therefore, the tank contained 100% air on the commencement of the loading. Approximately 3 m^3 had been loaded into No.2 cargo tank when the explosion occurred. Evidence points to the origin of the explosion being less than 1.4 m from the bottom of the tank (see photographs). The sudden initial flow rate into tank No.2 may have produced an aerosol mist. It was concluded that the formation of an aerosol with subsequent charging of the liquid particles was the most probable cause of the static discharge.

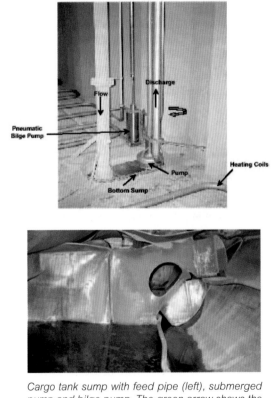

Cargo tank sump with feed pipe (left), submerged pump and bilge pump. The green arrow shows the supposed site of the explosion (between the wall and discharge pipe).

To prevent the production of a mist (splash filling) inside a tank or a static discharge, it has been good practice to restrict the filling velocity in the pipeline to 1 ms^{-1} until the outlet of the fill line has been covered by a minimum depth of 0.5 m. It is better to avoid going into the flammable region of vapour/air mixture by first inerting the tank to remove the oxygen content to below 5%. *Note: Benzene has a flash point of 12°F (−11°C), a flammable range of 1.4 to 8.0% and low conductivity of 5 x 10^{-3} pS m^{-1}. It is therefore a static accumulator requiring special precautionary measures to avoid a potential ignition risk. It is also a known carcinogen.*

For more details on loading/unloading procedures for ships, barges, rail or road cars, including bottom loading and on sampling and dipping procedures, refer to BP Process Safety Booklet *Tank farm and (un)loading safe operations.*

Miscellaneous

Storage tank filling hazards are similar to those listed for tank cars, trucks, tankers and barges. However, tanks are sometimes provided with slotted wells so that gauging can be done during filling. The well eliminates the chance of a spark discharge between the liquid surface and the end of a gauging rod or bob.

Filling portable containers is similar to filling tank vehicles except that the degree of hazard is reduced. When filling metal containers, the fill spout or nozzle should *always* be in contact with the edge of the container opening, unless a bonding wire is used.

The use of belt-driven equipment in hazardous areas should be avoided if at all possible because belts generate static electricity. This static electricity can be controlled by using belts made with conductive rubber, but periodic replacement is required because the belts become less conductive with use.

Heavy static charges accumulate on sandblasting hoses and hose fittings unless the hose is shielded with a metallic braid or wrapping that is bonded to the fittings and connected to a ground.

Because steel tanks and piping are themselves adequate grounds, it is seldom necessary to provide special grounding for them.

Stay away from steam clouds in hazardous areas; otherwise, you may become charged with static electricity, and the spark discharge from you to a grounded object may start a fire or explosion. Wet steam or steam containing rust particles will generate a static charge when exhausting to atmosphere. Any conductive object insulated from earth (you with rubber-soled shoes or a tank truck on rubber tyres are good examples) can become charged if near the steam jet, as shown in Figure 21a.

Figure 21a In rubber-soled shoes, you can get quite a 'charge' from just being near a steam jet—so stay away. Otherwise, the spark discharge from you to a grounded object might cause an explosion or fire.

Figure 21b Refinery operator testing his shoes at the beginning of his shift for an anti-static conductivity between 10 and 50 ohms.

Further, it is not necessary to be in contact with the steam cloud; just being close is enough. The general rule is to keep ungrounded conductive objects away from steam jets.

It must be recognized that there is no single 'cure-all' for the static electricity problem, but continued intensive study and investigation may lead to improved means of protection.

There is ample documentation that describes safety precautions to be taken to prevent electrostatic discharge hazards. A short list of these safety publications follows:

- International Safety Guide for Tankers and Terminals (ISGOTT), latest edition
- American Petroleum Institute Recommended Practice 2003 (API 2003) 'Protection against ignitions arising out of static, lightning, and stray currents'
- NFPA 77 'Recommended practice on static electricity'
- American Waterways Shipyard Conference 'Safety guidelines for tank vessel cleaning facilities'
- Tanker Safety Guide, Chemicals, International Chamber of Shipping, third edition, 2002
- BP Process Safety Booklet *Tank farm and (un)loading safe operations.*

Remember that other products used in petrochemical plants are static generators and accumulators (steam, water, CO_2). The BP Process Safety Booklets *Hazards of Water* and *Hazard of Steam* give examples of static electricity incidents generated by water, fire-fighting foams or steam.

A number of serious accidents occurred when very large crude carriers (VLCC) first came into service in the 1970s. Water washing techniques then in use caused the generation of large static charges in the ships' large cargo tanks. Often, ship and barge compartments were cleaned with high-velocity sprays or jets of water. This cleaning process may produce static charges, or the water jets may dislodge the magnesium anodes sometimes used for corrosion protection. When the anodes fall, they may strike a rusty surface and produce sparks hot enough to cause ignition. To be safe, be sure the compartment is free of any flammable mixture (either by gas freeing or purging with an inert gas) before water spraying begins. Figures 19 and 20 show the damage resulting from an explosion which occurred in a ship compartment during such a cleaning process.

The problem was largely eliminated by the use of crude oil washing (COW) and vessel inert gas systems (IGS).

Figure 19 Deck view of ship damaged by an explosion which occurred in a cargo tank being cleaned with high-velocity jets of water. Red arrows indicate where pieces of deck plate landed.

Figure 20 Side view of the ship shown in Figure 19. The ship was damaged by an explosion which occurred in a cargo tank being cleaned with high velocity jets of water.

4.3 The legend of mobile phones

In the recent years, an increasing numbers of reports have been accusing mobile phones of being the source of ignition of hydrocarbon vapours in service stations (and at least one on an offshore platform). BP carried out its own initial research and also asked a specialized contractor firm to study cases. Taking two different approaches both experts concluded that there was no possibility that petroleum vapours could be ignited on a retail forecourt from the normal use or carrying of such a device. This has also been confirmed by other specialists at the 2003 Institute of Energy (UK) conference on this subject (papers available by contacting the IE).

In fact, after obtaining the full details of the incidents, the root cause could be traced to static build-up on the person fuelling the vehicle while entering and exiting the car during the fuelling operation with the latch engaged on the fuelling nozzle. A number of states in the USA (including parts of Florida and New York states) and most countries in Europe have banned the latching practice.

Pumps and car damaged after a static electricity discharge ignited gasoline vapours in a US service station.

To summarize the studies:

- Mobile phones can be considered safe in the particular circumstances of filling a petrol (gasoline/diesel) car tank;
- However, mobiles phones still may present the following hazards:
  - distraction of the user;
  - by being possible sources of ignition in more flammable atmospheres (such as hydrogen, LPG, dust clouds) that have a minimum ignition energy level a lot lower than gasoline;
  - by perturbing electronic equipment (there are many incidents recorded where a 2 Watts explosion-proof rated radio activated a plant emergency shutdown system because it was used near an electronic gauge which then gave a false reading during the radio transmission).

Therefore if it can safely be considered that mobile phones are not an ignition hazard when filling a petrol tank in a service station, they are still to be strictly controlled and managed on petrochemical sites (depot, refineries, platforms), and the distraction issue is to be considered.

5

Hazards of lightning

Lightning is a frequent hazard to structures and the storage of flammable liquids.

A typical lightning stroke (see Figure 22 on page 30) may produce power of the order of a thousand billion horsepower, but only for a few millionths of a second. This high rate of energy release gives lightning its destructive effect.

A tremendous voltage (about 100,000 volts per foot of gap) is required to cause lightning to jump the gap between clouds or between cloud and earth. Sometimes storm clouds passing over a refinery may not develop enough voltage between cloud and earth to cause a lightning stroke. However, the voltage between the cloud and earth may be great enough to cause small electrical discharges (called *brush discharges* or 'Saint Elmo's fire') from sharp points on high vessels and structures. The burning of flammable mixtures coming from vents, even without a stroke of lightning, indicates that these brush discharges can cause ignition.

Direct strokes of lightning on a cone roof storage tank containing flammable liquids can ignite the tank's contents. The tank can be protected by riveting, welding or otherwise metallically connecting (bonding) the internal supporting members to the tank roof at not more than 10 ft intervals. The tank can also be protected by overhead ground wires or masts to divert direct strokes from the roof, if experience justifies the need for such protection.

Floating roof tanks with hangers in the vapour spaces of the seal are usually ignited as an indirect result of lightning rather than by a direct stroke. These ignitions generally result from bound charges that accumulate on the roofs of the tanks. Correlations between rimseal fire frequency and thunderstorm frequency have been developed in the Lastfire study (available on *www.resprotint.co.uk*). Typical frequency for Northern Europe sites is 1×10^{-3}/tank year; 2×10^{-3}/tank year for Southern Europe, North America and Singapore; and up to 13×10^{-3}/tank year in Venezuela or Thailand; and 21×10^{-3}/tank year in Nigeria.

Therefore, a refinery having 50 large floating-roof tanks in the US or South Europe has statistically one rimseal fire every 10 years (with possible escalation): $50 \times 2 \times 10^{-3} = 0.1$ fire/year $= >1$ fire/10 years.

This crude oil floating roof tank rim seal fire was started by lightning.

Bound charges are charges that are drawn up from earth into the tank roofs by charged clouds passing overhead during a thunderstorm. They return to earth slowly and harmlessly if the cloud passes on its way without producing a stroke of lightning. On the other hand, the bound charges are suddenly released if the cloud is discharged by a lightning stroke. This happens even though lightning does not hit the tanks but strikes in the vicinity. There is a case on record where six floating-roof tanks were ignited at the same time when lightning struck a radio tower outside the tank field.

The sudden release of bound charges produces large currents which rush to ground through the hanger linkages and shoes of the seal.

Poor electrical contacts or sharp points on the linkages cause the sparks which are the source of ignition.

The following measures can help minimize the effects of lightning on floating-roof tanks with a vapour space in the seal:

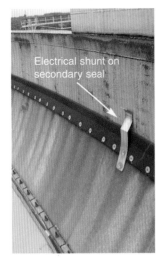

- Bond the roof to the shoes of the seal at intervals not greater than 10 feet on the circumference of the tank. For tanks with secondary seals, the bond should be from the roof to the shell above the secondary seal.

- Break up conducting paths through the hanger linkages by inserting insulating sections.

- Install a short jumper around each pinned hanger joint.

- Cover sharp points on the hangers with insulating material.

Floating-roof tanks without a vapour space in the seal (such as foam seals) are bonded between the roof and shell in a manner which will not damage the soft seal envelope.

Flexible cables connecting the floating roof to the rim of the tank shell will not provide lightning protection. Lightning will not follow long looped cables but instead will take a direct path through the hangers and shoes to the shell of the tank.

Figure 22 Picture of a floating roof to shell shunt test (submitted to a 830 A current to simulate lightning) showing the sparks generation (note that wax and rust deposits increase sparking).

It is very likely that such sparking will ignite any vapour present near the seal area, therefore the importance of seal integrity.

Picture from tests by Culham Electromagnetics and Lightning Limited for the Energy Institute (UK) and the American Petroleum Institute.

Special grounding of steel tanks is not needed for lightning protection. Tanks resting on the earth, or even on concrete rings with piping disconnected, have such a low electrical resistance to ground that special grounding is not necessary.

Experience indicates that above ground pressure vessels containing LPG or other flammable liquids under pressure do not have a lightning problem.

Structures may be protected against direct lightning strokes by rods or masts. These rods or masts provide a cone of protection within which equipment is immune to direct strokes of lightning. The radius of the protecting cone at the base is equal to the height of the mast (Figure 23).

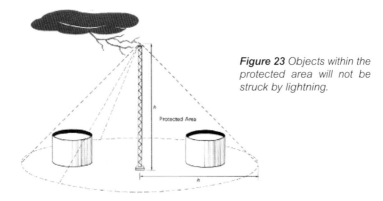

Figure 23 Objects within the protected area will not be struck by lightning.

Important electrical lines are often protected from lightning by an overhead ground wire which is called a *static wire* (Figure 24). The overhead ground wire provides a zone of protection, within which the electrical lines are shielded from direct strokes of lightning.

Figure 24 An overhead ground wire, sometimes called a static wire, protects these power lines from lightning.

There is almost no personal hazard from lightning if the following rules are observed:

- Do not go outside or stay outside during thunderstorms unless it is absolutely necessary. Stay inside a dry building, away from stoves and other metal objects.
- If there is a choice of shelter from a thunderstorm, choose in the following order:
  o large buildings made of metal or having metal frames;
  o Buildings protected against lightning by rods, masts, etc;
  o large buildings which are not protected;
  o smaller buildings which are not protected by lightning rods, masts, etc.

31

- If you must stay outside during a thunderstorm, seek shelter in an automobile (Figure 25), cave, depression in the ground, deep valley or canyon, thick woods or grove of trees, or at the bottom of a steep, overhanging cliff. By all means stay away from the following:
  - small sheds or shelters which stand alone in open spaces;
  - isolated trees;
  - wire fences;
  - wide-open spaces;
  - hill tops;
  - towers, columns, tops of structures and roofs of tanks.

When seeking protection from lightning in an automobile, remember that you must stay *completely inside*. If you attempt to get out, you risk being struck. Similarly, if you find yourself in a car with fallen electrical wires on or around the car, stay completely inside. Any attempt to leave the car under such circumstances may result in your electrocution.

Courtesy of Westinghouse Electric Corp.

Figure 25 *Demonstration showing that lightning (in this case, a 3 million-volt stroke from high-voltage generator) will not harm a person inside an automobile. Note the current flowing to ground near the right front tyre.*

6

Hazards of stray currents

Stray currents (those which are not intentional) flow through piping and connected vessels which are in contact with the ground. These currents result from power-line leakage or from the battery action of different kinds of soils or metals.

Two undesirable results from stray currents are (1) corrosion and eventual destruction of underground piping or other metallic objects (Figure 26), and (2) fires or explosions which may result from an arc which can occur when contacts (such as pipe flanges) are parted.

Current flowing

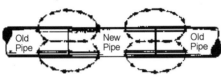

Figure 26 Even if the new pipe is of the same material as the old, the fact that the new is clean and the old is rusty is enough difference to produce current and corrosion of the new pipe.

Corrosion of an underground metallic object occurs at the point at which stray direct current flows from the object. The remedy lies in forcing direct current to flow to the object (cathodic protection) at that point. This can be accomplished by burying a bar of magnesium nearby or by providing a source of direct current, such as a rectifier.

The greatest danger from stray current results when pipe which is in a gas or light-oil service is disconnected. As the flanges are parted, an arc may occur which will be hot enough to ignite hydrocarbons present in the pipe or in the pit in which the pipe is located. If the presence of hydrocarbons is known (by a gas test) or suspected, and if there is a possibility that stray current is flowing in the pipe, the following precautions must be observed:

- Have fire-control equipment standing by for personnel protection.
- Shut off any cathodic-protection rectifiers in the area where the work is to be done. This stops the flow of cathodic-protection current through the pipes. Do not dig in the vicinity of a cathodic-protection system unless someone familiar with the system is present.

- Before parting a flange or cutting a line, install a bonding cable (jumper) across the point in the piping system to be opened (Figure 27).

- Do not disconnect the bonding cables at the pipe because an arc may occur at that point. Disconnect the bonding cable first at a lug a safe distance away, or leave the cable connected until the pipe is again made up by installation of the new fitting or section.

- When reinstalling a valve, spool or pipe section, bonding cables should again be used (Figure 27).

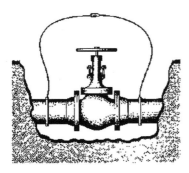

Figure 27 Procedure for removal and replacement of valve or spool when hydrocarbons and stray currents may be present (same procedure to be used for any break in line):
• Attach bonding cables.
• Remove valve or spool.
• Bonding cables may be removed, breaking connection first at lug. Cables must be reinstalled, making lug connection last before new valve or spool is installed.

Tank-car loading or unloading racks can be a source of danger because of stray currents in associated pipelines and/or spur tracks. Bonding (connecting with a piece of wire or metal) the tracks to the pipelines (Figure 28) is used to eliminate arcing. Insulated joints in both the tracks and pipelines entering and leaving the racks may be used to minimize the flow of stray currents. In addition, grounding the tracks (Figure 28) provides an effective ground for the tank car, and no special ground connection to the car is needed.

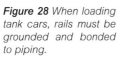

Figure 28 When loading tank cars, rails must be grounded and bonded to piping.

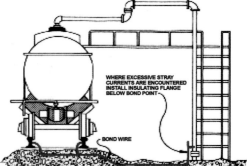

WHERE EXCESSIVE STRAY CURRENTS ARE ENCOUNTERED INSTALL INSULATING FLANGE BELOW BOND POINT

BOND WIRE

Ship-to-shore bonding wire

ISGOTT (International Safety Guide for Oil Tankers & Terminals) does not recommend a bonding wire between ship and shore. The bonding wire has no relevance to electrostatic charging. Its purpose was to attempt to short circuit the ship/shore electrolytic cathodic protection systems so that currents in hoses and metal arms would be negligible. It has been found to be quite ineffective and a possible hazard to safety. Insulation flanges or a single length of non-conducting hose is recommended to prevent the flow of current between the ship and shore. Refer to ISGOTT for more details.

Insulating flange signs (to prevent inadvertent removal or painting).

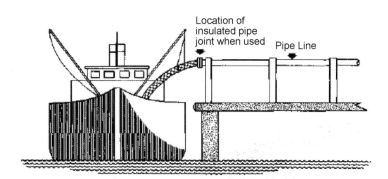

Figure 29 *Typical insulating flange position for ship loading.*

Note: Although the potential dangers of using a ship/shore bonding cable are widely recognized, attention is drawn to the fact that some national and local regulations may still require a bonding cable to be connected. If a bonding cable is insisted upon, it should first be inspected to see that it is mechanically and electrically sound. The connection point for the cable should be well clear of the manifold area. There should always be a switch on the jetty in series with the bonding cable and of a type suitable for use in a Zone 1 hazardous area. It is important to ensure that the switch is always in the 'off' position before connecting or disconnecting the cable. Only when the cable is properly fixed and in good contact with the ship should the switch be closed. The cable should be attached before the cargo hoses are connected and removed only after the hoses have been disconnected.

Example of poor practice: applying the bonding cable on a painted handrail will not provide good contact with the ship metallic body.

Welding machines are a dangerous source of stray current whenever the ground return cables of the machines are not directly attached to the work (Figure 30).

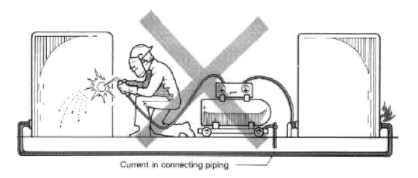

Current in connecting piping

Figure 30 *A welding-machine ground return cable which is not attached directly to the work can cause fire.*

The practice of attaching the ground return cable to any handy piping or steel structure may be convenient, but the unknown return path of the current can result in sparks or arcs at unexpected places. Fires can occur (and have occurred) because of this careless practice. Welding machine ground return cables must be attached directly to the object on which welding is being done (Figure 31). The same precautions apply when using welding machines for thawing frozen lines.

Figure 31 *The ground return cable for welding machines must be attached to the object on which welding is being done.*

7

Bonding and grounding

Bonding and grounding are essential to electrical safety, so it is important to know what the terms mean. *Bonding* means connecting two objects together with metal, usually a piece of copper wire. *Grounding* consists of connecting an object to earth with metal, and again a piece of copper wire is generally used.

The connection to earth is usually made to a ground rod or underground water piping.

7.1 For dissipation of static charges

Metallic objects charged by static electricity can always be discharged by connecting the charged objects to ground or to another metallic object which in turn is connected to ground. Tank-truck loading is typical of this situation because a tank truck is easily charged by filling. Spark discharge of the static electricity on the truck during loading can result in fire or explosion.

Always connect the grounding cable to eliminate this hazard.

Bonding a charged metallic object to a second metallic object insulated from earth will not necessarily do away with the charge. It will only cause the two objects to share the original charge of the first object, with a reduction in voltage between the objects and ground. The amount of this reduction depends upon the size of the second object. The larger the second object, the greater the reduction of voltage. The earth, being the largest body available, is very effective, reducing voltages to practically zero.

This is illustrated in Figures 32 to 35. Figure 32 shows two metallic bodies insulated from earth; one body is charged and the second body is uncharged.

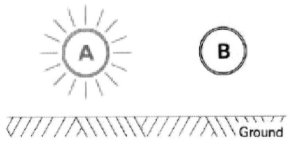

Figure 32 A charged body 'A' and an uncharged body 'B', both insulated from earth.

Figure 33 shows the charged body 'A' bonded to the second body 'B'. Both bodies now share the charge. The total charge remains, but the voltage to ground is reduced. For example, if body 'B' is the same size as body 'A', the voltage to ground will be halved.

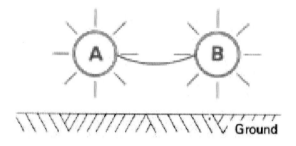

Figure 33 Metallic bodies share charge when bonded together.

Figures 34 and 35 illustrate the point that to effectively disperse charge on objects insulated from ground, it is necessary to bond them directly to ground or to other objects that are already grounded.

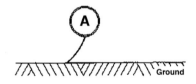

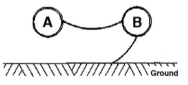

Figure 34 A metallic body connected to ground has no charge.

Figure 35 A metallic body connected to a grounded metallic body has no charge.

7.2 For protection of personnel and equipment

Electrical equipment is grounded for protecting personnel and also for protecting equipment.

Electricity, like gas or liquid in a high-pressure pipeline, is always looking for a way out. It can be contained by insulation (Figure 36), but if the insulation fails, the electricity seeks to flow to ground. If you are part of that path to ground, you get an electrical shock (Figure 37).

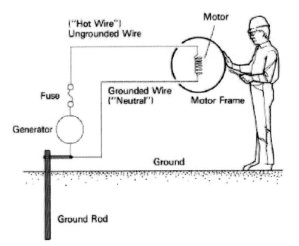

Figure 36 *Only good insulation protects you from shock if electrical equipment is not grounded.*

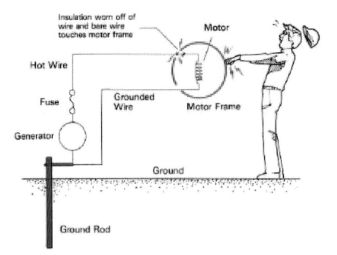

Figure 37 *Ungrounded electrical equipment is a shock hazard when insulation breaks down.*

The whole purpose of grounding for personnel protection is to provide electricity with a very easy metallic path to ground. Electricity always looks for the easy way out and if an easy path (one of low resistance) is provided, it will not flow through the body (Figure 38).

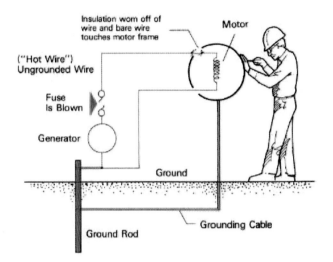

Figure 38 Properly grounded electrical equipment is free of shock hazard.

Most fixed equipment (pump motors, switch-racks, and the like) is grounded by metallic connection to ground rods or water pipes. Extreme caution should be observed when using portable equipment (welding machines, power tools, etc.), to be certain that the equipment is grounded (Figure 39).

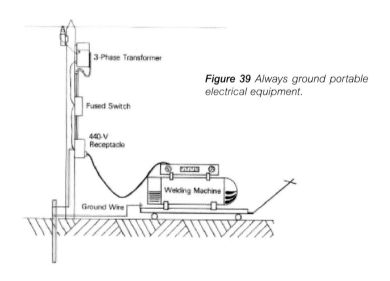

Figure 39 Always ground portable electrical equipment.

The user should connect the equipment to ground (water piping is excellent) through use of appropriate cable and clamps. Small power tools should have a three-prong plug, with the third prong used for grounding.

The only safe rule to follow regarding portable electrical equipment is this—do not make the power connection unless you are certain that the equipment is grounded, or will be grounded when the connection is made, as with a three-prong plug and a third wire for grounding.

Proper grounding of portable tools and electrical appliances at home is just as important as it is at work. The first step in electrical safety at home is providing a good ground connection to the local water system with a suitable jumper around the water meter (Figure 40). This jumper bypasses the meter with its nonconducting gaskets and ensures that water piping anywhere in your home will be an excellent ground.

Small 120-volt portable tools should be grounded with a three-prong plug and three-wire cord. Washers, dryers and electric stoves should be grounded directly to water piping or with a three-prong plug and three-wire cord (Figures 41 and 42). Do not forget, however, to make the necessary connection at the appliance terminal block if you want to ground a 220-volt appliance terminal through the three-wire cord and three prong plug. It is *absolutely* vital to safety that all electrical equipment in a damp basement be properly grounded, as the use of underground equipment in a damp location is an open invitation to electrocution.

Figure 40 Grounding jumper around water meter.

Figure 42 Three-prong plug and three-conductor cord used for appliance grounding (continental Europe & UK type with fuse).

Figure 41 Electrical appliances grounded directly to water piping.

In addition to preventing a hazard to personnel, grounding also protects equipment. An adequate ground (assuming a grounded system) permits a large enough quantity of short-circuit current to flow in order to trip the protecting circuit breaker or blow the protecting fuse when insulation fails (Figure 43). Otherwise, equipment might burn or be severely damaged because of a high resistance to ground.

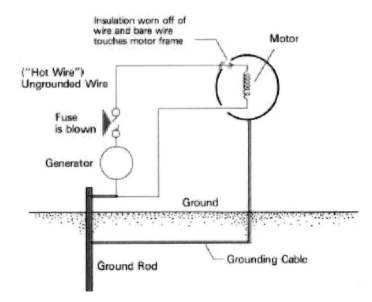

Figure 43 Insulation failure will cause the protective fuse to blow if the electrical system and equipment are properly grounded.

ACCIDENT Figures 44 and 45 show a barrel house which burned because a short length of conduit was not adequately grounded. Wiring insulation in the length of conduit failed, and current flowing through the high-resistance path to ground generated enough heat to start the fire.

Proper grounding would have prevented all this by causing the protecting fuse to blow when the insulation failed.

Figure 44 Barrel-house fire caused by poor electrical gounding.

Figure 45 Proper grounding would have prevented this fire.

8

Hazards of electrical shock

Approximately 1,000 persons in the US die each year as a result of electrical shock. Direct death from electrical shock results from ventricular fibrillation, paralysis of the respiratory centre or a combination of the two. *Ventricular fibrillation* is a condition wherein the heart quivers but does not beat. It is caused when a certain amount of current passes through the heart area. Paralysis of the respiratory centre occurs when a certain amount of current passes through the respiratory control centre in the brain. In such a case, breathing stops and death results from asphyxiation.

The amount of current and the current's path are two important factors affecting the extent of injury. The amount of current depends on voltage and body resistance. Body resistance can be very high or very low, depending on whether the skin is dry (high resistance) or wet (low resistance). Contact area also affects body resistance—a person in a bathtub has both wet skin and a large contact area and is almost certain to be electrocuted if a shock is received. Current path through the body will be directly from the point of entry to the point of exit, because current flows through all parts of the body with equal ease. Electrocution may occur when the heart area or the respiratory control centre of the brain is in the current's path.

The amount of current which the human body can tolerate is very small, and is measured in milliamperes (a *milliampere* is one thousandth of an ampere). A rough guide based on the *Accident Prevention Manual for Industrial Operations*, published by the National Safety Council, is as follows:

Milliamperes	Current injury
1 to 8	Shock sensation
8 to 15	Painful shock
15 to 20	Painful shock with control of adjacent muscles lost. Individual can not let go.
20 to 50	Painful shock with severe muscular contractions and difficult breathing.
50 or more	May be fatal

Under special conditions (say, wet skin and, hence, low resistance) voltages from 45 to 60 have proved fatal. Any voltage above 12 is considered dangerous and becomes more dangerous as it increases.

A few simple rules should be observed to prevent electrical shock or to minimize injuries resulting from electrical shock.

8.1 Rules for prevention of electrical shock

- Be certain that all electrical equipment with which you must work is properly grounded or disconnected from the source of power.

- Be particularly careful if you are physically tired, as that is when the greatest number of accidental electrocutions occurs.

- At home, keep electrical devices (radio, heater or hair dryer, for example) out of the reach of anybody in the bathtub. Electrocution may occur when the person in the tub touches the device or when the device falls into the water.

Remember also to ground household appliances and power tools as discussed in Chapter 7.

Always use the adequate safety devices and wear your Personal Protective Equipment (PPE). It can save your life!

ACCIDENT An electrician was badly burned as a result of an electrical flash/explosion that occurred while he attempted to install two 277/480 volt self-contained meters. The electrician removed the glass covering the empty sockets, visually inspected the sockets, verified the fuses were open on the load-side, checked voltage of both sockets, and the grounding on the load-side of both sockets. He then proceeded with the installation. He carefully inserted the bottom clips of the meter into the socket. He rotated the meter up so that the top clips would make contact, at which time the meter would have been fully inserted into the socket.

The moment the top clips made contact, the electrical flash occurred. The fault blew the fuse in the 2500 kVA transformer. The electrician sustained burns in the face and spent two days in hospital. He was wearing all the PPE required by the procedure for this work.

Burnt PPE.

Burnt electrical panel.

The investigation team concluded that the test procedures were being properly followed and that the meter was installed correctly. No exact cause has been determined, but the explosion was believed to be due to a fault in the electrical switchgear. Additional PPE (face shield) is now required for undertaking such electrical work.

> Wear protective equipment for the eyes or face wherever there is danger of injury to the eyes or face from electric arcs or flashes or from flying objects resulting from electrical explosion.

8.2 Rules for minimizing injury from electrical shock

- Cut the power supply.
- If cutting the electricity supply is not possible, try to free the victim of electrical shock from contact with the live conductor at once, using a dry stick, dry rope, dry clothing or other nonconductor. *Do not touch him with your bare hands!* (Figure 46)

Figure 46 Free victim from contact with live conductor at once, using dry clothing, rope or stick. Don't touch him with your bare hands!

The life of a shock victim may be in danger because of little or no breathing. In such a case, begin rescue breathing at once, as follows (Figure 47):

1. Tilt the head back with victim on his back, neck fully extended.

2. Elevate victim's jaw into jutting-out position by inserting thumb between teeth, grasping lower jaw and lifting it forcefully upward.

3. If air passage is not yet cleared, clear at once with several sharp blows between the shoulder blades.

4. Open your mouth wide and cover victim's mouth completely by placing your mouth over his with airtight contact, also closing victim's nose by pinching it between thumb and finger.

5. Blow air into victim's lungs until you see the chest rise (less forcefully for children); remove your mouth and let him exhale. If chest does not rise, check steps above.

6. Repeat step 5 approximately 12 times a minute until victim revives (20 times a minute for children).

7. In the case of infants, rescue breathing should be done through both the nose and mouth.

8. If the victim's mouth can't be opened, rescue breathing should be applied mouth-to-nose, rather than mouth-to-mouth.

- Never give an unconscious victim any liquids.

- Efforts to aid a victim should be continued until he revives or is pronounced dead by a physician. A victim of electrical shock often appears dead, with no pulse, heartbeat or breathing, but these appearances are often deceiving and can best be interpreted by a physician.

- Also remember that even if electrical burns look minor at the entry and exit points, there might be invisible internal damage. Consult a doctor immediately after an electrical shock.

The above rules for reviving shock victims cover some of the more important points, but they are not intended to serve as a complete guide. You can obtain complete information from your first aid trainer or such organizations as the National Safety Council or the Red Cross.

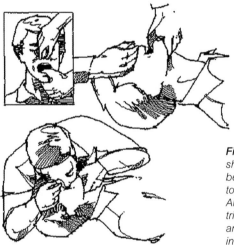

Figure 47 If a victim of electrical shock has little or no breathing, begin mouth-to-mouth or mouth-to-nose rescue breathing at once. Also remember that even if electrical burns look minor at the entry and exit points, there might be invisible internal damage. Consult a doctor immediately after an electrical shock.

9

Explosion and fire hazards of electrical equipment

Explosion-proof electrical equipment presents no explosion hazard if properly installed and maintained. It is designed to withstand the pressure created by an internal explosion and to cool hot gases below ignition temperature before they reach the outside of the explosion-proof housing, as shown in Figure 48.

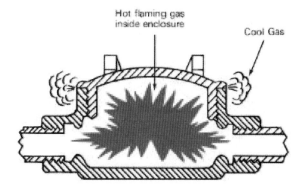

Figure 48 *Explosion within an enclosure.*

This cooling is accomplished by routing the escaping gas between closely machined flanges (Figure 49) or threaded joints (Figure 50).

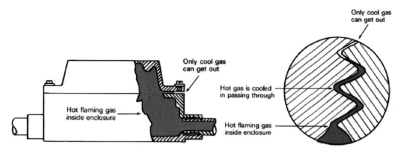

Figure 49 *Enlarged view showing a normal flat joint.*

Figure 50 *Enlarged view showing a normal threaded joint.*

The primary danger results from carelessness—flange faces may be scratched or dirty; cover bolts may not be tightened; threaded covers may not be fully engaged. In such cases, the gases escaping from an internal explosion may still be hot enough to ignite flammable mixtures outside the explosion-proof housing.

For safety, observe the following rules when installing or maintaining explosion-proof equipment:

- Flange surfaces must be clean and undamaged.
- Threads must be clean and undamaged.
- When replacing covers, tighten bolts properly or engage all threads on threaded covers.
- If the above three conditions cannot be met, replace the damaged equipment.

The table below show different types of equipment (and relevant European norms):

Concept	Symbol	Icon	Description	Category *	EN Standard
General requirements					50014
Oil immersion	Ex o		explosive gas excluded by immersing ignition source in oil	2	50015
Pressurized	Ex p		explosive gas excluded by surrounding ignition source with pressurized inert gas	2	50016
Powder filled	Ex q		explosive gas excluded by immersing ignition source in sand	2	50017
Flameproof	Ex d		ignition within the apparatus enclosure is contained and will not ignite surrounding explosive atmosphere	2	50018
Increased safety	Ex e		Design excludes the possibility of incendive arcs, sparks or hot surfaces	2	50019
Intrinsic safety	Ex ia		energy in circuit and temperature on components reduced to a safe level	1	50020
	Ex ib			2	
Non-incendive	Ex n		will not ignite explosive gas in normal operation, faults unlikely to occur	3	50021
Encapsulation	Ex m		flammable gas excluded by encapsulating the ignition source in resin	2	50028

See next page.

Learn how to read the labels!

Certified equipment for use in potentially explosive atmospheres must be labelled adequately. For example, in Europe, equipment must be labelled as follows:

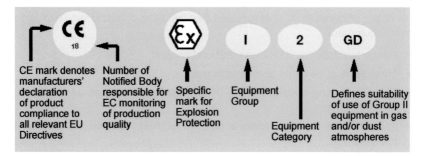

Equipment group:

I = mining
II = industrial (other than mining)

Equipment category:

1 = 'very high protection'; can be used in zones 0, 1, 2, 20, 21, 22.
2 = 'high protection'; can be used in zones 1, 2, 21, 22.
3 = 'normal protection'; can be used in zones 2, 22.

The certification code is explained below:

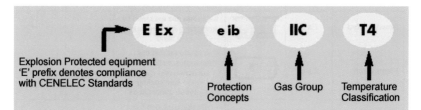

Gas Group	Product	Minimum ignition energy
I	Methane	280 micro Joules
IIA	Propane	250 micro Joules
IIB	Ethylene	70 micro Joules
IIC	Hydrogen/Acetylene	11/20 micro Joules
II	All gases	

Temperature classification (maximum temperature of equipment surface):

T1 < 450°C (842°F)   T4 < 135°C (275°F)   *See appendix 3 for*
                                                examples of ignition
T2 < 300°C (572°F)   T5 < 100°C (212°F)   *temperatures for*
                                                common products
T3 < 200°C (392°F)   T6 < 85°C (185°F)

Do not take a chance. Extreme caution must be observed when non-explosion proof electrical equipment located in a nonhazardous area is connected to conduits which run underground or to a hazardous area. Flammable gases or vapours sometimes get into the conduit and follow it back to the non-explosion proof equipment. This happens in spite of the fact that conduit seals are placed between the source of the gas or vapour and the equipment. *Conduit seals cannot be relied on to keep flammable gases or vapours out of electrical equipment.*

ACCIDENT This has been demonstrated several times when vapourtight lighting panelboard housings in non-hazardous areas exploded as panelboard switches were operated (Figures 51 and 52). The result was death in one instance and injury in the others. This is not a condemnation of the vapourtight equipment itself, since it was not designed to contain an explosion.

However, it does illustrate that care is required in selecting locations where it may be used.

Figure 51 Gas entered vapourtight lighting panelboard housing through underground conduits and exploded when a panelboard switch was operated.

Figure 52 Explosions inside vapourtight lighting panelboard housings shattered the covers and caused injury and death.

Investigation of these accidents revealed that flammable gases or vapours had travelled through underground conduits, passed through conduit seals of the type approved by Underwriters' Laboratories, Inc., and entered the panelboard housing (Figure 53). If there is a chance that flammable gases and vapours can get into an electrical conduit, switches and other arcing devices connected to that conduit must be explosion-proof.

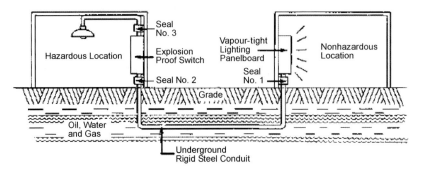

Figure 53 *Ineffective conduit seal No. 1 permits gas from leaking underground conduit to enter the vapourtight (cast metal) lighting panelboard where explosion causes housing to shatter.*

Figure 54 shows a close-up view of a vapourtight lighting panelboard similar to those that exploded. Figure 55 shows a close-up view of an explosion-proof lighting panel board which is designed to contain an internal explosion.

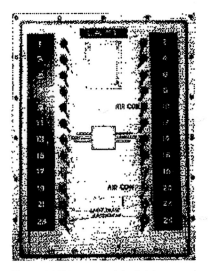

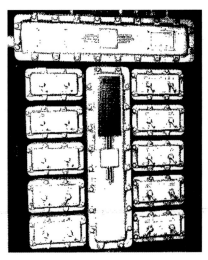

Figure 54 *This vapourtight lighting panel-board is typical of those shattered by the explosion.*

Figure 55 *This explosion-proof lighting panel board is designed to contain an internal explosion.*

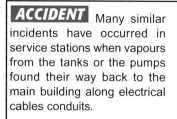 **ACCIDENT** Many similar incidents have occurred in service stations when vapours from the tanks or the pumps found their way back to the main building along electrical cables conduits.

For certain usages in hazardous areas, it is impossible or impractical to obtain explosion-proof equipment. Certain instruments and large synchronous motors are examples. In such cases, the housing for the individual item of equipment may be pressured. If sufficient equipment is involved, it may be more practical to pressurize the room in which the equipment is located (Figure 56). As an example, process unit switchrooms and control rooms are sometimes pressurized. Pressurizing consists of keeping the equipment housing or room under positive pressure with uncontaminated air.

Thus, the uncontaminated air can leak out, but no gases or vapours can leak in.

Needless to say, if a pressurizing system fails, an explosion hazard may result. To avoid any hazard, observe the following rules:

- Keep the pressurizing system (instrument air for instrument housings, motor-driven fans for large synchronous motors and rooms) functioning properly.
- Keep the housing or room closed so that a positive pressure will be maintained. It is impossible to pressurize a room, for example, if windows or doors are kept open. Thus, most control rooms and some switchgear rooms are air-conditioned to aid in maintaining a closed space.
- Hydrocarbon samples in open containers should not be brought into a pressurized room.

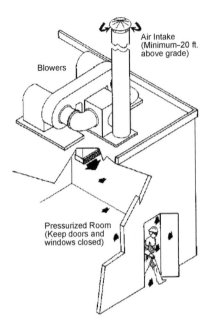

Air Intake
(Minimum–20 ft.
above grade)

Blowers

Pressurized Room
(Keep doors and
windows closed)

Figure 56 A pressurized room is kept under positive pressure with uncontaminated air. The uncontaminated air can leak out, but gases or vapours cannot leak in.

Drains and breathers are provided to help keep explosion-proof enclosures free of moisture. Because metal-to-metal contact is required to cool gases escaping from an internal explosion, explosion-proof enclosures never use gaskets. Consequently, moisture condenses inside the enclosure as a result of breathing caused by temperature changes. Drains permit the moisture to escape, and breathers permit free passage of air to help keep the interior dry. An occasional check should be made to see that breathers and drains are not obstructed.

In some instances, a small electric heater is installed in an enclosure to help keep it dry.

9.1 Batteries

Another explosion hazard exists with batteries as they can discharge hydrogen to the atmosphere during normal operation. Hydrogen is a very light gas and rises rapidly in air. It has a very wide flammability range (4–74%) in air compared with other hydrocarbons gases/vapours (2–15%). A hydrogen mixture requires very little energy for ignition.

ACCIDENT There were hydrogen explosions in the battery rooms of two buildings, resulting in extensive damage.

The first explosion blew a 400 sq ft (37 m²) hole in the roof, in addition to collapsing numerous walls and ceilings throughout the building. The company had vacated the building and moved out the computer equipment, leaving the battery back-up system behind. Three days prior to the incident, a local alarm was reported but it was not relayed to the Fire Department. It appeared that batteries were charging for a long period without any ventilation. The hydrogen built up and then found an ignition source.

The second similar explosion ripped off a 40 ft (12 m) section of the roof. A computer firm handling data collection had occupied the building a month prior to the incident. The battery backup system was suspected to be the cause of the explosion.

Both buildings happened to be vacant at the time and there were no injuries.

Incident 1   Incident 2

Lessons learned

- Local alarms are no good if there is no one there to respond to them.
- Whenever a flammable substance is present in a confined area, the release of that material can create conditions for an explosion.
- In the case of battery rooms, the presence of hydrogen must be recognized and a good, fail-safe ventilation system is essential to dilute the released hydrogen to maintain a concentration below the lower flammable limit, taking into account the H_2 generation rate.
- Decommissioning of the computer/data centre should have included the isolation of the battery charging system.

Large banks of batteries are not necessary for hydrogen explosions. Look at the photographs below, where single batteries have exploded (one in a car engine compartment). The biggest risk for vehicle batteries is during battery charging in garages or in confined spaces for prolonged periods such as overnight! Have you ever left a car battery charging in a confined space?

9.2 Electrical fires

Many serious electrical substation fires occur regularly in petrochemical plants and refineries, some of which resulted in large consequential losses.

ACCIDENT After a styrene plant had experienced a power supply malfunction and the DCS alarm had activated, a technician was asked to investigate the cause in the switchroom located on the third floor of the building.

He discovered smoke in the stairwell, alerted the shift supervisor who then called the site's fire brigade by telephone.

The fire brigade found a fire in a cabinet when they entered the switchroom. Equipped with breathing apparatus, they extinguished the fire with CO_2. There were no injuries. The fire was unrelated to the power supply malfunction/alarm which was associated with another cabinet.

The exact technical cause of the incident could not be determined because of the damage done to the cabinet in the fire.

Extent of damage to cabinet.

9.3 Fire protection of electrical/instrumentation/ computer rooms

Incipient fire detection (IFD) using a laser smoke detector system provides a very early sign of incipient combustion. It works by continuously monitoring the room by drawing samples into a detection chamber where the air sample is exposed to a laser light source. An output signal is generated dependent upon the smoke density. This signal can provide a local alarm or be relayed back to the control room.

All personnel who may respond to an IFD system require training. Unlike the traditional smoke detector where there will be visible smoke if there is a fire (unless it is a false alarm), with an IFD system there will invariably be no smell, smoke or flames.

No fixed fire suppression system is necessary with IFD. It is recommended that $CO_2 \times 5$ kg hand extinguishers be provided adjacent to all compartments protected by IFD. IFD is ideally suited for electrical switchgear, servers, computers, control rooms, telecommunications, electrical substations etc.

Previously, compartments both on and offshore would have both fire detection and a halon 1301 extinguishing system. However, with IFD, no fixed fire suppression is normally necessary. Where a detailed risk assessment has determined the benefits for installing a fixed fire suppression system, then either a fire water mist/spray/fog or CO_2 system would be recommended depending upon the specific situation.

Appendix 1 also details recommended practices for diesel driven emergency generators and pumps, which are too often the cause of incidents by lack of attention to design, operation and maintenance.

ACCIDENT

Explosion of a switchgear cubicle caused by the introduction of a foreign body which caused a sudden short-circuit. The flashover occurred, initially between two phases and later escalated to 3-phase to earth, burning the technician on the face and hands. Local smoke detection (IFD) operated as a result of the flashover.

Cable overheating detected by IFD well before a fire occurred.

10

Dangers of improper operation of electrical equipment

Improper operation of electrical equipment may damage the equipment itself, create a hazardous situation, or both.

Motors and motor starters are probably the most abused of all items of electrical equipment. The irony of this is that most of the abuse comes from those who are going out of their way to be kind to the equipment. The well meaning souls who 'bump' a motor onto the line (*bumping* consists of alternately operating the start and stop button so that a motor does not accelerate smoothly) are in reality doing the motor and starter a great disservice (Figure 57).

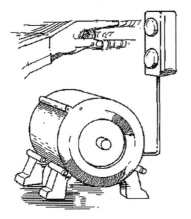

Figure 57 Frequent starts and stops shorten motor life.

While starting and coming up to speed, motors take much more than normal full-load current. Repeated application of this large current is damaging to both motor and starter and shortens life. When starting a motor, push the start button (just once). Let the motor and starter do the rest without interference.

Sometimes electrical equipment is used in a dangerous manner even though it is not being abused. Often, in a hazardous area, someone will remove a guard, globe and light bulb and screw in a plug to obtain 220 or 120-volt power even though an approved type of receptacle is close by (Figure 58).

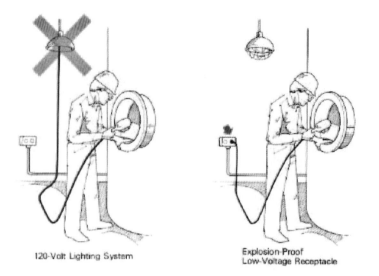

120-Volt Lighting System

Explosion-Proof
Low-Voltage Receptacle

Figure 58 Using a light-bulb socket as a source of power is always a shock hazard and can also be an explosion hazard in a refinery.

Such poor practice, of course, constitutes an explosion hazard and a shock hazard. Explosion-proof plugs and receptacles should always be used in hazardous locations, and all 220 or 120-volt plugs and receptacles should be of the grounding type regardless of location.

It is important for personal safety to use extension lamps operating at 12 volts or less when going inside a metal enclosure such as a tank or a vessel (Figure 58).

Equivalent personal protection is provided by 220 or 120-volt portable lights equipped with ground-fault circuit interrupters, as the GFI will trip the circuit when current (4 to 6 milliamps) flows from the protected circuit to ground. Do not use this system where flammable liquids and/or vapours may be present.

Another important safety rule in connection with extension lamps is that they must be properly guarded to prevent breakage. The hot filament from any broken bulb can cause ignition, and the resulting darkness and broken glass are also hazards.

Simple things such as warning lights can be misused. In one instance, someone was about to string warning lights (bare bulbs and open wiring) around a leak in a propane line! Fortunately, he was stopped in time.

Even in the office or at home, never overload or misuse electrical circuits and equipments (see two pictures below).

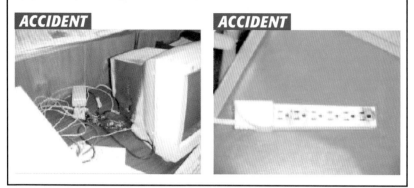

Rules for the safe use of electrical equipment are as follows:

- Do not deliberately overload electrical equipment or circuits.
- Do not bump motors.
- Do not use improper equipment in hazardous areas. If you don't know what is safe and proper, ask someone who does. If you are still in doubt, get a gas test to make sure the work area is hazard-free.

11

Safeguards for electrical equipment

11.1 General

Safeguards for electrical equipment are designed to protect people from injury, to prevent people from damaging or misoperating equipment, and to protect the equipment from damage by natural causes.

Often electrical equipment is located in locked rooms or fenced enclosures. Respect those rooms and enclosures by staying out unless you are authorized to enter (Figure 59).

Figure 59 Do not enter switch-rooms or fenced enclosures around electrical equipment unless you are authorized to do so. Once inside, always follow instructions.

Besides risking personal injury, unauthorized persons may damage equipment and interrupt production.

Most switchrooms or fenced enclosures are designed with two exits. If your work requires you to enter such an area, know where all the exits are and be sure they are free and clear for rapid use.

ACCIDENT In one instance, a painter working in a compressor room not only violated instructions to stay out of the fenced electrical equipment area, but he used a ladder to get over the fence because the door was locked. While painting the ceiling near an open-type switch, he was electrocuted.

Figure 60 A man disobeyed orders and climbed a switchroom fence to paint a ceiling. He touched and was electrocuted by a switch like these shown here.

Pushbutton guards (Figure 65) prevent accidental operation of motors and serve also as a reminder that motors are to be started and stopped only by authorized personnel. Stay away from pushbuttons, whether they are guarded or not (Figure 66). A large single motor (11,000 horsepower) used on a refinery process unit was shut down by a painter who accidentally bumped a pushbutton, resulting in shutdown of the process unit.

Figure 65 Pushbutton guard.

Figure 66 Give pushbutton stations a wide berth. Accidental starts and stops are dangerous.

11.2 Contact with overhead cables

In most areas, there is no mechanical safeguard for protection of overhead electrical lines from crane booms. However, in refineries and chemical plants, it is good practice to install warning girders over roads before electrical lines or pipe-racks.

ACCIDENT

Good practice: Warning girder on both sides of a pipe-rack (note that the fact that the first girder looks higher than the rack is an illusion due to the picture being taken from ground level).

What can happen when pipelines are unprotected... Luckily, in this case, no leak occurred despite a natural gas line being impacted.

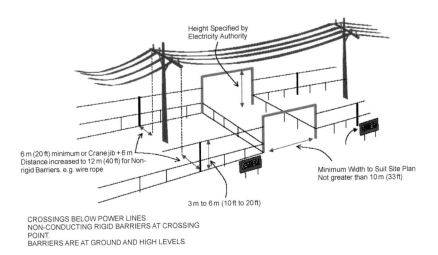

Height Specified by
Electricity Authority

6 m (20 ft) minimum or Crane jib + 6 m
Distance increased to 12 m (40 ft) for Non-
rigid Barriers. e.g. wire rope

Minimum Width to Suit Site Plan
Not greater than 10 m (33 ft)

3 m to 6 m (10 ft to 20 ft)

CROSSINGS BELOW POWER LINES.
NON-CONDUCTING RIGID BARRIERS AT CROSSING
POINT.
BARRIERS ARE AT GROUND AND HIGH LEVELS.

The best safeguard is a careful crane operator. Too often, overhead lines have been knocked down, resulting in lost production and fire hazard. More seriously, falling wires are a hazard to personnel, and wires lying on the ground present a serious shock hazard.

ACCIDENT What can happen if lifting operations are not planned correctly? This crane boom touched a high voltage line. The crane's hydraulic fluid and tyres ignited when the current found its way to the ground.

When cranes operate near overhead lines, someone should stand by with the primary duty of warning the crane operator when the crane boom gets as close as permissible to the overhead lines (Figure 67).

Figure 67 Someone should have the primary duty of directing crane operation near overhead electrical lines.

ACCIDENT A very serious accident occurred because nobody watched the clearance between a crane boom and overhead electrical lines. A pipefitter, standing on some pipes, attempted to put a sling on a crane hook. The hook was not in the proper position, so the pipefitter took hold of it and signalled the crane operator to raise the boom as necessary. The boom came in contact with high-voltage electrical wires, and the pipefitter, who completed the electrical circuit to ground as shown in Figure 68, required mouth-to-nose rescue breathing and was burned on 40 percent of his body. *When a crane boom is moving near overhead electrical lines, don't grab the hook or touch the crane!* Wait until the boom has stopped and you know there is safe clearance between the boom and the electrical wires.

Figure 68 When a crane boom is moving near overhead electrical lines, don't grab the hook or touch the crane! Wait until you are sure the boom has stopped a safe distance from the wires.

Sometimes crane booms accidentally come in contact with electrical wires. When this happens, observe the following rules:

- If you are in the cab or on the crane, stay there! You are as safe as a bird on a wire or the man in the car which is being struck by lightning. Do not get off the crane until the operator has moved the boom a safe distance from the wires. If this cannot be done, stay on the crane until an electrician has shut off the current and shorted and grounded the wires.

- If you are walking in the area, stay clear! You may be electrocuted by touching any part of the crane or by coming in contact with fallen wires.

These rules apply whether the crane is mounted on rubber tyres, steel treads or rails. Special coverings over the insulation on wires are provided to prevent damage. Some may be of the armored type to prevent mechanical damage; some may be thermoplastic or neoprene to prevent corrosion and mechanical damage; and some may be lead to prevent entrance of moisture.

ACCIDENT A five-man crew was erecting warning 'goal posts' on either side of a railway equipped with overhead electrical power lines. They carried a metal pole under the 3000–3500 VDC high voltage electrical power line and held it vertically, contacting the power line. A worker suffered a fatal electric shock, and another received burns to his hands.

Railway crossing. Metal pole contacted the overhead power line located on the right hand side of the photo. The truck crossing the railway line provides scale. The overhead power line is approximately 5.7 m above the ground. The metal pole measured 6.0 m in length.

The following pictures are typical situations of contact with overhead lines extracted from incidents reports:

ACCIDENT

ACCIDENT

Repaired Line

Original Line Position

Original Truck Position (crane boom was extended at the time of accident)

ACCIDENT Before a lifting operation, also think that it can go wrong and that anything under the path of the lift or the cranes can be impacted (pipelines, electric cables…) — see picture below as an example in which two electric cables were cut.

When leaded cables are installed underground, it is sometimes necessary to use cathodic protection to prevent corrosion of the lead. In the case of overhead lines, a weatherproof covering is used to protect the copper conductor. This covering is not insulation, and therefore an overhead wire should be treated as if it were bare as far as shock hazard is concerned.

11.3 Excavation work

Lack of planning and rigour in excavation work is also a major cause of electrocution and power outages. The incident description is a typical example.

ACCIDENT A project for a new plant required a new 11 kV feed from an existing substation. The work involved sub-sub-subcontractors. Prior to starting the excavation, the area was assessed, cable scans carried out, and the route agreed (this was between two groups of existing cables).

Once trenching work was completed, the cable laying contractor observed a section of damaged cable in the trench wall but did not report it. An earth fault occurred on the existing 33 kV feeder between two substations 20 days after the end of the trench work (circuit breaker failed to operate because it had been disabled at some point in the past). The fault was cleared in a third substation.

The resulting power loss caused major disruption across the petrochemical complex.

Causes:

- During the excavation work, pneumatic clayspades were used at depths below 300 mm, despite clear instruction to use only hand tools below this depth. At some stage during the excavation, a clayspade damaged the 33 kV feeder, resulting in a subsequent earth fault.
- At some point in the past, the circuit breaker had been disabled by two sections of plastic cable ties inserted between the relay and its current transformer, probably during performance testing of the relays.

Lessons learned:

- Planning and execution of power supply and distribution work requires thorough risk assessment, must involve appropriate competent resource, and must be subject to sufficient supervision to ensure that control measures and work methods are applied.
- Utility systems such as power distribution systems have the potential to cause major incidents.

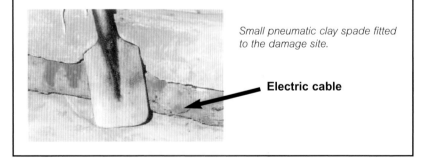

Small pneumatic clay spade fitted to the damage site.

Electric cable

11.4 Training and safe work practices

Special training is required for work on electrical equipment. Only authorized employees may conduct electrical work. This training for authorized employees covers:

- safe work practices;
- isolation of electrical sources;
- test equipment;
- tools and PPE.

Specialized energy isolation measures are an essential part of plant safety and must be considered as an integral part of the safe working practices and procedures, such as permit-to-work (for more detail on this, refer to BP Process Safety Booklet *Tank farm and (un)loading safe operations*).

Before starting work:

- de-energize, lock, tag and test all circuits;
- de-energize all power sources;
- disconnect from all electric energy sources,

The main power disconnect device (usually located in the switchgear room) should be turned to the OFF or OUT position. Control circuit devices such as push buttons, selector switches, interlocks, etc may not be used as the sole means for de-energizing circuits or equipment.

To lock and tag all sources means that a lock and a tag must be placed on each disconnecting means used to de-energize circuits:

- attach lock so as to prevent operating the disconnecting means;
- place a 'Do Not Use' or 'Do Not Operate' tag (Figure 69) with each lock.

Figure 69 Typical 'Do Not Use' tag.

This main disconnect device should be padlocked and a 'Repair Hold Card' should be signed, dated and attached (Figure 70).

REPAIR HOLD CARD

Figure 70 Typical 'Repair Hold Card'.

The local control switch should be checked again to make sure the equipment is 'off'. Note that for a pump motor equipped with an automatic start, it may not be possible to prove the integrity of the isolation by attempting to start the electric motor from the local start button.

In any event, local safety regulations covering such situations should always be strictly followed. Equipment which operates from or contains more than one source of voltage should be provided with proper warning signs.

ACCIDENT A contractor received injuries to his thumb when a pump started automatically while he was greasing the coupling. When the contractor came to work on that pump, he checked its isolation status by operating the stop/go button as indicated on the permit. Since it did not start, he assumed that isolation had been carried out and commenced work. However, the pump was remotely controlled by the pit level so the stop/start button was not active when the pump was in auto mode. During the maintenance operation, the pit filled and the level switch started the pump, causing injury to the contractor's thumb.

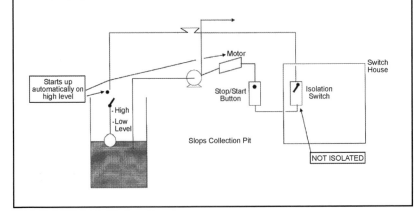

ACCIDENT Contractors were carrying out the cleaning of air coolers. In order to accomplish this, they had to open a trap in the cage located below the blades of each air cooler. The contractors were aware that each trap had a padlock with a specific key but that some padlocks were in bad condition and might have to be cut because the key would not open the padlock. A work permit was required for each air cooler. Electrical isolation of the motor of the air cooler was carried out.

As indicated on the work permit, a contract worker set out to clean air cooler A, after receiving the key to that air cooler from the operator. However, the contractor went to the air cooler B.

When the key didn't work the padlock, the contractor thought that it was due to the bad condition of the padlock. He cut the padlock and entered the cage.

Fortunately, an operator saw the contractor in cage B and stopped the work.

Actions resulting from this incident included a requirement for an operator to be present when the trap of a cage of an air cooler has to be opened, and a requirement that the fan motor of the air cooler be tested before authorizing entry into the cage (to be sure the motor is well isolated).

Some plants may use different methods, such as the multiple padlock procedure wherein each person (or craft) working on the equipment in question puts his own padlock on the main disconnecting device and signs it in and out. This is a good practice when there are multiple tasks (crafts) involved on a single equipment.

If a lock cannot be applied, a tag used without a lock must be supplemented by at least one additional safety measure that provides a level of safety equal to that of a lock (supervision approval required).

Examples of this situation include:

- removal of an isolating circuit element such as a fuse;
- blocking of a controlling switch;
- opening of an extra disconnecting device.

Stored electric energy must be released before starting work:

- discharge all capacitors;
- short-circuit and ground all high capacitance elements.

Verify system is de-energized:

- operate the equipment and control to check that equipment cannot be restarted;
- use test equipment to test the circuits and electrical parts for voltage and current (check test equipment (Volt-Ohlm Meter) on a known live source of the same rating to ensure it works before and after checking the circuit on which you will be working).

11.5 Identification of disconnecting means and circuits

Each disconnecting means for motors and appliances shall be legibly marked to indicate its purpose. Each service, feeder, and branch circuit, at its disconnecting means or over current device, must be legibly marked to indicate its purpose.

ACCIDENT An electrician received an electric shock from 20 kV current while he was working on a live electrode of the desalter of a distillation unit. The electrical equipment used in the desalter was characterized by the fact that only one pull-out circuit breaker was used to protect the three transformers, each of which supplied one electrode grid. In order to continue operating the desalter with the two remaining electrodes, another electrician had disconnected the electrical supply circuit to the faulty electrode's transformer (#3) on the circuit breaker side in the substation.

The electrician went to work on electrode #1 because there was a lack of identification on the three transformers, and because the mimic panel available in the control room showed the transformers in the order 1, 2, 3, when in fact they were in the order 3, 2, 1 when looking from the main access route.

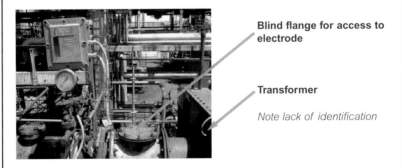

Blind flange for access to electrode

Transformer

Note lack of identification

Fortunately, he was not electrocuted due to the fact that his hand was in contact at that moment with the equipment which was earthed/grounded when the tool that he was holding touched the live cable. He suffered severe burns on his little finger and third finger of his right hand.

The investigation concluded that the main contributing factors were as follows:

- inadequate checks/tests to determine that the electrical conductor to the electrode was dead;
- inadequate identification/tagging/labelling of the electrical equipment.

The fact that there was only one circuit breaker for three transformers was also listed as a design anomally but it did not contribute directly to the incident.

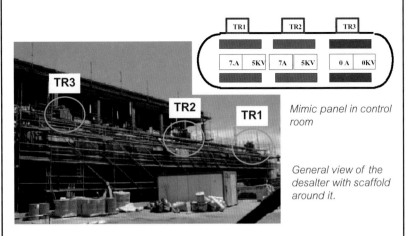

Mimic panel in control room

General view of the desalter with scaffold around it.

Note: Mimic panels are graphical representations for control purposes. Layout drawing and equipment databanks must be referenced and such information kept up-to-date.

11.6 Re-energizing equipment

When the equipment has been repaired, a responsible person must verify that it is ready to return to service:

- conduct tests and visual inspections to ensure all tools, electrical jumpers, shorts, grounds, and other such devices have been removed;
- warn others to stay clear of circuits and equipment;
- each lock and tag must be removed by the person who applied it;
- visually check that all employees are clear of the circuits and equipment.

The 'Repair Hold Card' should be signed off and removed, as well as the padlock. The 'Do Not Use' tag at the local control switch should also be removed.

11.7 Working with energized equipment

Sometimes electricians must work on live circuits or apparatus. On such occasions, rubber gloves and blankets, 'hot sticks', and other special devices are used. Such equipment must be in perfect condition and used in strict accordance with safety rules. Persons working on energized equipment must be familiar with the proper use of special precautionary techniques, personal protective equipment, insulating and shielding materials, and insulated tools.

When working on energized circuits:

- isolate the area from all traffic;
- post signs and barricades;
- use an attendant if necessary;
- use insulated tools, mats and sheeting;
- use electrical rubber sheeting to cover nearby exposed circuits.

Do not work on energized electrical parts:

- without adequate illumination;
- if there is an obstruction that prevents you seeing your work area;
- if you must reach blindly into areas which may contain energized parts.

ACCIDENT An electrician was working on a de-energized cabinet with a box spanner. He did not realize that when turning the spanner, its long end reached into the cabinet below, where live wires were present. The electrician suffered severe burns and was saved only by the prompt reaction of his colleague who shutdown the substation power supply.

Use only non-conductive tools and proper safe work practices when required to work on live (hot) electrical equipment.

ACCIDENT An electrician was using a metallic mirror to inspect the back side of a breaker connection in a 480 Volt panel when a metal part of the mirror came into contact with an energized portion of the breaker, causing a large electrical arc. Two employees received second degree burns from the flash over. Two process units had to shut down because of the power surge.

It is also important to make sure that discarded equipment is made safe, and if at all possible, that it is fully dismantled, including its energy and control sources. The incident below is a good example of a 'booby-trap' left unintentionally when taking an equipment out of service:

ACCIDENT A 5000 barrel crude tank was being cleaned when an explosion lifted it several feet off the ground, splitting the roof open ⅓ to ½ the circumference at the roof seam and shooting a yellow flame horizontally 20 to 30 feet (6–9 m) out of the roof opening. The vapours coming from an open hatch ignited on the 300V DC line that was left seven years before when an ultrasonic level sensor was dismantled.

- Decommissioning of ultrasonic level sensor was not properly or completely performed.
- Lockout-tagout procedures for energy isolation were not followed.

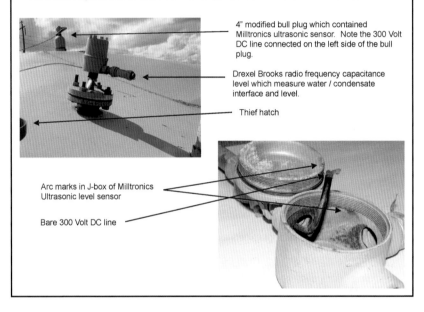

4" modified bull plug which contained Milltronics ultrasonic sensor. Note the 300 Volt DC line connected on the left side of the bull plug.

Drexel Brooks radio frequency capacitance level which measure water / condensate interface and level.

Thief hatch

Arc marks in J-box of Milltronics Ultrasonic level sensor

Bare 300 Volt DC line

12

Power outages

12.1 General

Power outages are defined as an interruption or perturbation of a common energy supply (usually electricity (85% of cases) but also steam (13%), gas, fuel, hot oil, etc). They can be 'partial' (for example, loss of power to a solenoid activating an ESD valve) or 'total' (such as loss of all electricity supply to all process units).

The expected consequences of power outages are:

- massive flaring;
- loss of utilities (electricity, air, cooling water, steam, etc);
- stress on equipment (thermal cycle, vibrations, water hammer, etc);
- stress on personnel ('christmas tree' alarming, unexpected event, multiple manual tasks both on shutdown and start-up, etc);
- short downtime and start-up costs.

Unfortunately, some power outages also have more serious outcomes:

- equipment rupture (piping, rotating equipment, etc);
- overfilling (flare system, vessels, etc);
- hydrocarbon/chemical releases;
- fire/explosion;
- injuries;
- water/ground pollution;
- overwhelming of personnel/systems;
- external consequences;
- public outrage;
- repair and inspection costs;
- long downtime costs;
- other power outages by domino effect between close-by plants through local grid.

From a 2003 study on 23 serious power outages, the average cost of a refinery outage is estimated at 2,334 k$.

It must be considered that more than 20% of incidents occur during shutdown or start-up and therefore, an unexpected shutdown and associated start-up means extra stress on equipment and personnel (refer to the BP Process Safety Booklet *Safe Ups and Downs for Process Units*). A typical example is described below.

ACCIDENT Before the incident, there had been severe thunderstorms passing over the refinery, causing power interruptions and lightning strikes. The strikes caused a fire in the Crude Distillation Unit, and resulted in a shutdown of the FCCU and many other units.

Process unit heat balances became upset, resulting in the loss of some facilities. At the process control room, numerous alarms were triggered and information on process parameters was affected, to the point that some of it became misleading.

While attempting to re-start the FCCU later that morning a debutanizer relieved and filled a flare drum with liquid. The wet gas compressor tripped with the result that the FCCU main fractionator overhead gas make was discharged to flare. This gas flow, in combination with liquid from the overfilled flare drum—i.e., two phase flow—created mechanical shocks and vibration in the flare header piping. The 30-inch diameter pipe at the flare drum outlet ruptured at its weakest point (the sudden hydraulic force of the hydrocarbon liquids entering the flare line caused it to break at the elbow bend).

Approximately 20 tons of flammable hydrocarbons escaped to atmosphere from the outlet pipe of the flare knock-out drum on the fluidized catalytic cracking unit (FCCU). The drifting cloud of vapour and droplets ignited about 110 m from the flare drum outlet and the force of the explosion was equivalent to 4 tons of high explosive. This was followed by a fire.

Picture courtesy of the Western Mail and Echo Ltd.

The site suffered severe damage and glass damage occurred in a nearby town (3 km away). Twenty-six people suffered injuries on-site, none serious. Rebuilding the damaged refinery was estimated at $76 million and the company was fined $320,000 with $230,000 legal costs.

12.2 The regulator point of view

A UK study involving 100 chemical companies was carried out by the Authorities (Health & Safety Executive) during 1992/93 following a major power loss at a chemical complex on Humberside to 'identify the extent of the current awareness of power loss/surge issues at chemical companies and whether specific risk assessments had been carried out, back-up systems installed and maintenance issues identified'.

This study identified that although there was a general awareness, the perception of power loss/surge incidents amongst industry was that they were mainly related to quality, production and profit issues rather than having safety implications.

The report concluded that 'recent power loss incidents have highlighted that power loss incidents may result in significant safety risks. Under the requirements of the Major Accident Hazard legislation, it is important that industrial sites review power loss/surge issues.'

12.3 Triggering events

Triggering events of power outages are often external ones (such as lightning, external grid power dip) but also often could have been avoided by better procedures or maintenance (for example, testing wrong equipment, cutting cable during trenching work). Whatever the triggering event, power outages can be prevented if protective devices are adequately designed, tested and maintained; and good practices adhered too for operations and maintenance.

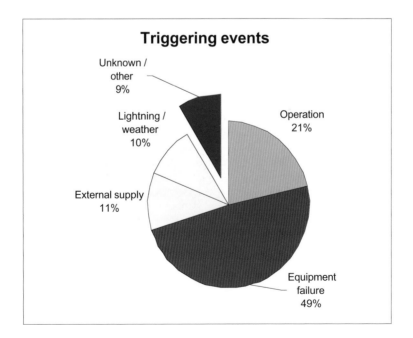

The usual sequence of events as described in incident reports is shown in the diagram below.

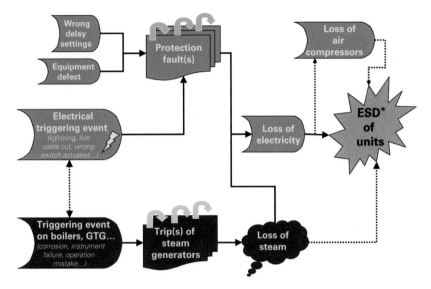

* ESD: Emergency Shut Down

ACCIDENT The refinery was shut down in a safe manner after having been isolated from the national electrical grid followed by shutdowns of the two Cogen units.

The incident occurred when re-energizing one of the six transformers that brings down the voltage from 25 kV (national grid) to 10 kV. The transformer had been out for overhaul that coincided with the Crude Unit turnaround.

A short circuit developed in the transformer that resulted in a voltage dip in the refinery. This in turn caused many electrical motors to trip after which the automatic restarting function came into operation. This drew a high current from the grid and because the refinery was exporting relatively much power, the total current through the breakers connecting the refinery grid with the national grid became higher than the trip setting. As a result the refinery was disconnected from the grid. The control system of the two Cogen units could not cope with such a rapid change from high exports of power to 'island operation'. This led to tripping of both gas turbines and hence a full shut off of power supply to the refinery.

The defect in the transfomer was caused by gradual build up of carbon on one of the collectors. This turned out to be a problem known to the supplier but was never communicated to end users.

12.4 What can be done to keep plants safe in case of a power outage?

Research—make sure you know this...

- Find out which instruments and equipment are on the emergency power system and which ones are not. Know how to compensate for the controls that will be lost.

- Review the operating instructions of the emergency power system. Know how long it is expected to operate. If you have an emergency generator, make sure you have operating instructions, fuel and any other items needed for its operation.

- Know the power off fail position of key instruments and equipment. Most are probably designed to go to a 'fail-safe position', but for those that are not, you should have instructions on what to do. Also, some fail-safe ones are so critical that they may need to be checked in the field.

Preplan—think through what you would do...

- Do you know the things to do immediately if a utility stream (steam, electricity, instrument air) failed? Make sure that you know where to find the relevant emergency procedures quickly at all times for the next steps. If needed, carry them with you at all times using small/short copies.

- Know the calls and notifications that you need to make and know how to make them quickly.

- Review your emergency response plan and emergency procedures for actions to be taken.

- Accidents can occur during restart after a power outage. Know the safe restart procedure!

Practice—walk and talk through what needs to be done...

- Talk with your co-workers about what they would do and why. Develop a common plan for response to power failures. Know how to respond when all alarms sound at once. Report possible improvements of procedures to your supervisor.

- If you can, conduct or participate in power failure drills including the testing of emergency generators. If this is not possible, mentally walk through power failure actions.

> The sudden lack of electricity and other utilities can be shocking. Know your role in outages!

13

Some points to remember

1. Remember that very small arcs have enough energy to ignite a flammable mixture.

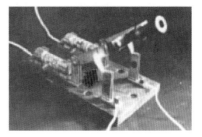

2. Those responsible for design and installation should know the National Electrical Code area classification, and be sure that all electrical equipment meets NEC requirements.

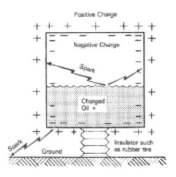

3. Be wary of static electricity. Remember that static electricity is generated whenever gasoline, kerosene, jet fuels and similar products are handled, particularly when loading at high velocities.

4. Ground tank trucks immediately after the truck is spotted for loading and before the loading spout is inserted. Remember that this eliminates the chance of a spark discharge from the tank truck to some external object, but it does not eliminate the chance of spark discharge from the oil surface inside the tank.

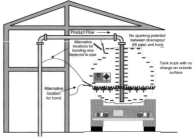

5. Beware of switch loading—the loading of a high flash point product (such as kerosene) after a load of low flash point product (such as gasoline) has been hauled. Remember that switch loading is the principal factor in loading-rack truck fires.

6. Remember that lightning or fallen electrical wires will not harm you if you stay completely inside your automobile.

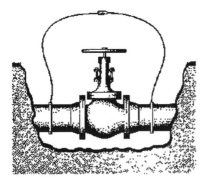

7. Use a bonding cable when cutting or opening a line, if hydrocarbons and stray currents may be present. Otherwise, an arc from stray current may provide a source of ignition.

8. When loading tank cars, be sure the rails are grounded and bonded to the fill piping. This eliminates the chance of a stray current arc between the fill pipe and tank car.

9. Attach welding-machine ground return cables to the object on which welding is being done.

10. Avoid shock by working only with grounded or disconnected electrical equipment. Be sure that portable electrical equipment is grounded.

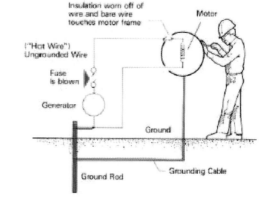

Insulation worn off of wire and bare wire touches motor frame

Motor

("Hot Wire") Ungrounded Wire

Fuse is blown

Generator

Ground

Ground Rod

Grounding Cable

11. Be particularly careful when you are physically tired, as that is when the greatest number of accidental electrocutions occur.

12. Free a victim of electrical shock from contact with the live conductor at once, using a dry stick, dry rope, dry clothing or other nonconductor. Do not touch him with your bare hands.

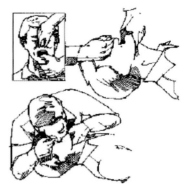

13. If a victim of electrical shock has little or no breathing, begin mouth-to-mouth or mouth-to-nose rescue breathing at once.

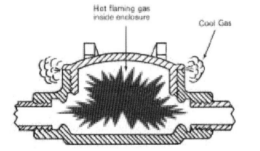

14. Keep the flanges or threads of explosion-proof enclosures clean and undamaged. When replacing covers, tighten bolts properly, or engage all threads on threaded covers.

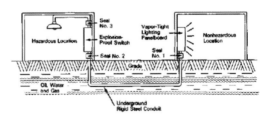

15. Do not trust a conduit seal to prevent the passage of flammable gas or vapour through a conduit.

16. Keep doors and windows of pressurized rooms closed.

17. Do not 'bump' motors. Push the start button just once, and let the motor accelerate smoothly and continuously to full speed.

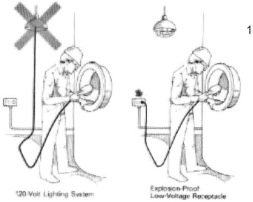

120-Volt Lighting System   Explosion-Proof Low-Voltage Receptacle

18. Use approved receptacles and plugs. Do not unscrew the nearest light bulb!

19. Use extension lamps operating at 12 volts or less when working inside tanks or vessels (120 Volt GFI circuits offer equivalent personal protection). Be sure the lamp is properly guarded to prevent breakage.

20. Stay away from pushbuttons to avoid starting or stopping motors accidentally.

21. When a crane boom is moving near overhead electrical lines, do not grab the hook or touch the crane! Wait until you are sure the boom has stopped a safe distance from the wires. If you are in the cab or on the crane when the boom hits electrical lines, stay there until the operator moves the boom a safe distance from the wires or until an electrician shuts off the current and shorts and grounds the wires. If you are walking in the area, stay clear! You may be electrocuted by touching the crane or coming in contact with fallen wires.

Appendix 1: Diesel driven emergency equipment

A1. Scope and introduction

From offshore platforms to refineries, depots, accommodation camps or research centres, fixed diesel driven pumps and generators are used in many installations, for either normal or emergency operations. From very small 2 V cylinders to huge 16 V or bigger, they often provide critical services such as fire water, emergency power, lighting.

Diesel electricity generator

Diesel fire pump

Such systems are often bought 'off the shelf' from manufacturers as a full kit and are not subject to a detailed engineering and HSE review, either before installation or during their life. If this type of installation is a good enough standard for temporary use, it is quite another matter when the same equipment is intended to be installed permanently, often being started remotely and functioning without human supervision for hours.

Many incidents, both inside and outside BP, have shown that simple low cost precautions could prevent most incidents.

ACCIDENT

Diesel generator destroyed by fire in a BP facility.

The scope of this note does not include marine engines, mobile or temporary installations, process pumps and fixed installations attended at all times when functioning. It also does not cover other HSE aspects such as noise or fumes emissions and does not replace relevant standards (such as NFPA 20) but is a complement where experience has shown that additional care was required.

A2. Engine equipment

Separation of diesel engines from hazardous areas should ideally be determined using a risk based approach. Some plant layout codes may prescribe a distance, such as 100 metres. Emergency equipment should in any case be protected from the effects of blast and/or thermal radiation from a fire. It is common to have multiple pump arrangements divided between two well separated pumphouses to protect against a common mode situation.

Both normal and backup pumps in same room.

Diesel pump

Electric pump

Both normal and backup pumps in same room.

The diesel fuel tank should be located outside the engine room. Most diesel driven engine skids have the fuel tank located under, or worse, above the engine.

Typical diesel driven emergency generator installed as a skid in a container to supply emergency power to a chemical plant.

Note:

- *diesel fuel tank on top of engine;*
- *no fire detection;*
- *no external stop button;*
- *no fire extinguisher nearby.*

Diesel generator in two hours fire rated room with fuel tank in adjacent room.

Note the lack of smoke detection.

Good separation of fuel tank from pump room but note:
- *no level gauge on fuel tank;*
- *no fire extinguisher.*

Despite being required by standards such as NFPA 20, many fuel tanks are not fitted with any level gauge other than a sight tube. Worse, this sight tube is often a flexible plastic hose that offers no fire resistance and often no UV resistance. Tanks should be equipped with a reliable level gauge that is both weather resistant and will not release the whole content of the tank if involved in a fire.

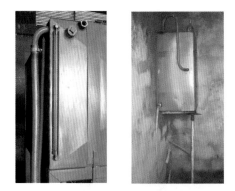

Two typical examples of diesel fuel tanks with a plastic gauge tube and plastic 'garden hose' type connections.

Similarly, fuel piping should not be made of 'garden hose' type flexible hoses. It should be possible to cut the fuel supply to the engine without entering the engine room.

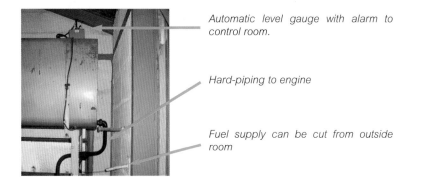

Automatic level gauge with alarm to control room.

Hard-piping to engine

Fuel supply can be cut from outside room

The tanks should be in a secondary containment that must provide storage of at least 110% of the tank's maximum capacity (refer to UK Environment Agency Pollution Prevention Guidance note 2, Feb 2004).

Example of unprotected diesel engine

Bad corrosion of fuel tank and no secondary containment for the full capacity

A3. Fire protection

Except in particularly sensitive applications/locations, no fixed fire suppression system is required for fixed diesel engines.

Example of a diesel fire pump in a container, fitted with a sprinkler system. In case of fire on the engine, stopping it would stop the water flow...

It should be possible to cut the fuel supply to the engine without entering the engine room. For electricity generators, it should be possible to isolate the generator from the grid from the outside of the generator room.

The engine room should have a smoke detection system and CO_2 portable fire extinguishers.

Example of diesel fire pumps with fuel tanks located far away.

Gasoline driven pump.

Note asbestos protection of flue gas duct

A4. Maintenance

All diesel engines should be tested at least weekly and should run for a minimum of 30 minutes each. Formal records should be made of each test and critical parameters, such as fuel levels in tanks and battery charge, should be noted.

All diesel engines should be serviced at least annually by a specialized contractor and overhauled once every ten years as a minimum.

Overhaul of a diesel fire pump.

Good housekeeping in engine rooms is required. All flammable materials (such as lubricant drums, oily rags) should be removed.

Appendix 2: Short bibliography for regulations and norms

US National Electrical Code (NEC) by the National Fire Protection Association (NFPA)

American Petroleum Institute Bulletin No. RP-500

OSHA standards:

- 29 CFR 1926.550(a)(15)(i) Clearance Between Electrical Power Lines and Cranes. OSHA STD 3-12.1A (9 May 1980),
- 1910.137, Electrical protective devices.
- Design safety standards for electrical systems:
  - 1910.302, Electric utilization systems.
  - 1910.303, General requirements.
  - 1910.304, Wiring design and protection.
  - 1910.305, Wiring methods, components, and equipment for general use.
  - 1910.306, Specific purpose equipment and installations.
  - 1910.307, Hazardous (classified) locations.
  - 1910.308, Special systems.

The Electricity at Work Regulations 1989 (UK)

The Dangerous Substances and Explosive Atmospheres Regulations 2002 (UK)

EU ATEX Directives:

- ATEX 95: European Union directive 1994/9/EC.
- ATEX 137: European Union directive 1999/92/EC.
- EN 1127-1: Explosive atmospheres—explosion prevention and protection. Part 1: Basic concepts and methodology.
- Non-electrical equipment for potentially explosive atmospheres:
  - EN 13463-1: Part 1: Basic method and requirements.
  - EN 13463-5: Part 5: Protection by constructional safety.
  - EN 13463-6: Part 6: Protection by control of ignition sources.
  - EN 13463-8: Part 8: Protection by liquid immersion.

- Electrical apparatus for explosive gas atmospheres:
  - ○ EN 60079-10: Part 10: Classification of hazardous areas.
  - ○ EN 60079-14: Part 14: Electrical installations in hazardous areas (other than mines).
  - ○ EN 60079-17: Part 17: Inspection and maintenance of electrical installations in hazardous areas (other than mines).
  - ○ EN 60079-19: Part 19: Repair and overhaul for apparatus used in explosive atmospheres (other than mines or explosives).

UK Environment Agency Pollution Prevention Evidence Note 2. February 2004.

Appendix 3:
Ignition temperatures for common dusts and gases/vapours

Dust Cloud	Ignition Temperature	
	°C	°F
Aluminium	590	1094
Coal dust (Lignite)	380	716
Flour	490	914
Grain dust	510	950
Methyl cellulose	420	788
Phenolic resin	530	986
PVC	700	1292
Soot	810	1490
Starch	460	860
Sugar	490	914

Note: Temperatures given here are the lowest ones found in different reference documents.

Gases/Vapours	Ignition Temperature	
	°C	°F
Acetone	465	869
Acetylene	305	581
Ammonia	615	1139
Benzene	498	928
Butane	287	549
Diesel oil	257	495
Diethyl ether	160	320
Ethane	472	882
Ethylene	425	797
Ethylene oxide	429	804
Furfural	316	601
Gasoline 50–60 oct.	280	536
Gasoline 100 oct.	456	853
Hydrogen	400	752
Kerosene	210	410
Methane	537	1000
Naphtha	290	554
Propane	450	842
Propylene	453	847
Styrene	490	914
Toluene	480	896
Trichloroethylene	410	770
White spirit	232	450

For more information, consult:

- for gases/vapours: IEC 60079-20;
- for dusts: BS 7535;
- and:
  - manufacturer's MSDS;
  - *Sources of ignition* by John Bond, ISBN 0 75061 180 4.

Test yourself!

1. Electricity cannot harm if there is less than 1 Ampere.

    True ☐   **False** ☐

2. Static sparks cannot ignite hydrocarbon vapours.

    True ☐   **False** ☐

3. Limiting the size of sampling container is a good way to prevent static accumulation.

    True ☐   **False** ☐

4. Hazard of filling containers can be evaluated as follows:

Greatest hazard	Ungrounded metal container
Less hazard	Non-conducting container (e.g., plastic container)
Least hazard	Grounded metal container

    True ☐   **False** ☐

5. The radius of the protecting cone of a lightning mast at the base is equal to the height of the mast.

    True ☐   **False** ☐

6. Insulation flanges or a single length of non-conducting hose is recommended to prevent the flow of current between the ship and shore.

    True ☐   **False** ☐

7. When charging, batteries emit hydrogen, which can create an explosive atmosphere.

    True ☐   **False** ☐

8. To work on a pump, after the main switch has been OFF and locked/tagged in the substation, the local control switch should be checked again to make sure the equipment is 'off'.

    True ☐   **False** ☐

9. Touching overhead electric cable is not hazardous as they are insulated.

    True ☐   **False** ☐

10. Power outages cannot be prevented.

    True ☐   **False** ☐

1F/2F/3T/4T/5T/6T/7T/8T/9F/10F

96

Your notes